101 TREES OF INDIANA

T0284042

MARION T. JACKSON

101

INDIANA UNIVERSITY PRESS

Illustrations by Katherine Harrington

NEW EDITION

TREES *of* INDIANA

Photographs by Ron Rathfon

This book is a publication of

Indiana University Press
Office of Scholarly Publishing
Herman B Wells Library 350
1320 East 10th Street
Bloomington, Indiana 47405 USA

iupress.org

Manufactured in China

Cataloging information is available from the Library of Congress.

First printing 2024.

978-0-253-06981-8 (paperback)

CONTENTS

Preface

By featuring nearly 150 species we suggest that Indiana's tree flora is approximately triple the number of species covered in the earlier popular booklet *Fifty Trees of Indiana,* which was long out of print, but is now again available. Although Deam and Shaw (1953), in *Trees of Indiana,* third edition, listed the total number of tree species native to our state as 101, they also included 13 species of introduced trees, along with more than 60 varieties, forms, and hybrids.

In this work, we feature those species native to Indiana, plus several that have been introduced and commonly encountered as street, lawn, or park ornamentals, or in commercial plantings. Several introduced species are now regenerating without human assistance; some are now aggressive invaders. Also included are a number of small trees that are shrubs in life form, as often as they reach actual tree size. Some species of these latter groups (both native and introduced) are illustrated and described less completely than are the native species that reach full tree size.

The featured species numerically represent the overwhelming majority of individual trees that occur in Indiana, plus they constitute essentially all of the vegetational matrix of natural forested communities of our state. Consequently, the odds are greatly in your favor that any given tree that you find in a natural forest, or other wild setting, will be among the 101 primary listees.

Additionally, that uneven numbers are believed in certain of the world religions to confer a karma or a sanctity to their adherents or users is, perhaps, a further justification for our selection of 101 featured species! May all users of this manual be blessed by acquiring a special knowledge of, and love for, the trees of Indiana. As author Willa Cather once wrote, "Today I stood taller from walking among the trees."

Acknowledgments

We are indebted to many for ideas, assistance, and encouragement to make the publication of this book a reality. While it is not possible to recognize everyone who contributes to a project such as this, and we realize that by naming some, we run the risk of omitting others who may be equally deserving, we wish to acknowledge a number of individuals and organizations.

Any note of recognition regarding the trees of Indiana must begin with Dr. Charles C. Deam, Indiana's first state forester, who authored the Indiana Classified Forest Act, originated the State Forest System, who knew the plants of Indiana more thoroughly than anyone before him or since, and who described them fully in his four volumes, the last of which was the co-authored 1953 edition of *Trees of Indiana,* published more than 50 years ago, when Deam was well into his eighties.

We acknowledge with gratitude, the many fieldworkers (particularly those with Indiana's Department of Natural Resources) who collectively accumulated much of the information about Indiana's trees and forestlands that has made this book possible. Dr. Burnell Fischer, his co-workers, and district foresters of the Division of Forestry, and John A. Bacone, Lee Casebere, and their staff of the Division of Nature Preserves, were especially helpful. Mike Homoya of the DNP gave invaluable advice regarding species to include in the "Species List of Indiana Trees."

New county distribution records were provided by many individuals, but the following merit special thanks: Brian Abrell, Kemuel Badger, Russell Brown, Cliff Chapman, Cloyce Hedge, Roger Hedge, Ron Helms, Hank Huffman, Rob Jean, Larry Jones, Ron Rathfon, Keith Ruble, Tom Swinford, Jim Wichman, and Indiana's district foresters of the IDNR.

William R. Overlease, a retired botanist of Bloomington, provided me with distribution maps for several exotic tree species now occurring in the state, and offered helpful comments on the book in general. John Siefert, now Director, Division of Forestry, DNR, suggested the book's functional shape.

Special mention is due to Bill Allen and Christi Hicks, both Life Sciences undergraduate majors at Indiana State University, who compiled much of the new county distribution record data. For the uncounted hours that both worked in the preparation of updated species range maps, we are truly grateful. The late Norman Cooprider, then staff cartographer at ISU, prepared the species distribution maps and graphs for the text.

The excellent photograph of Tulip-tree on page xi is from the vast collection of Ron Everhart of Indianapolis.

Funding for the inclusion of color photographs and to defray certain publication costs was provided by a contribution from the Wabash Valley Audubon Society and a special gift from Ms. Amy Mason of Terre Haute, Indiana. We are indeed grateful for this support of our project.

We are indebted to Indiana State University for the facilities and services provided, especially Ms. Laura Bakken, who very capably typed the manuscript and offered many suggestions for its improvement. Mss. Sharon Sklar, Bobbi Diehl, Linda Oblack, and Miki Bird of Indiana University Press coordinated the project and served as designer and successive sponsoring and copy editors.

John Bacone, Roger Hedge, and Michael Homoya of the Indiana Division of Nature Preserves, and Ron Rathfon, Purdue Extension Forester, critically reviewed the manuscript and offered many helpful comments. Any errors that remain are ours.

Respectfully,
Marion T. Jackson, Author
Katherine Harrington, Illustrator
Ron Rathfon, Photographer

INDIANA'S STATE TREE

The Tulip-tree—*Liriodendron tulipifera* L.

In 1923 the Indiana state legislature selected the flower of the tulip-tree as the state flower. But in 1931, the state legislature repealed the 1923 Act, and selected the tulip-tree as the state tree. The zinnia, a native of Mexico, then became the state flower.

No tree species could be more appropriate as Indiana's state tree. Native to nearly all areas of the state, ancient in lineage, majestic in form, impressive in dimensions, beautiful in all seasons, it truly graces the Indiana land-scape, either as a forest monarch, or as a handsome ornamental. It is a magnificent and valuable timber tree, as well.

Record tulip-trees in the original forest were truly amazing. The Indiana Geological Report of 1875 includes the following: "I measured four poplar trees that stood within a few feet of each other; the largest was thirty-eight feet in circumference three feet from the ground, one hundred and twenty feet high, and about sixty-five feet to the first limb. The others were, respectively, eighteen and a half, eighteen and seventeen feet in circumference at three feet from the ground."

Similarly, Wm. C. Bramble and Forest T. Miller reported in *Natural Features of Indiana* that "One tulip tree in the Wabash Valley measured one hundred and ninety feet tall, twenty-five feet in circumference at three feet above the ground, and ninety-one feet to the first limb" (1966).

Although none of those colossal monarchs of Indiana's original forest still exist, some very impressive individuals currently grace the Indiana landscape, many of which are protected in Indiana state parks and nature preserves. Two of the largest known individuals grow adjacent to each other in Hemmer Woods Nature Preserve in Gibson County (Jackson, 1969). Other fine forest-grown specimens may be observed in Donaldson's Woods in Spring Mill State Park, in Hoot Woods in Owen County, and at several other locations, where some individuals approach or exceed four feet in diameter, with clear boles of 60 or more feet. For more information about the tulip-tree, see Dr. Paul Rothrock's excellent essay in *The Natural Heritage of Indiana* (1997).

We are indeed pleased and proud that the tulip-tree or yellow-poplar became Indiana's state tree more than 70 years ago.

IN MEMORY OF THE AMERICAN CHESTNUT

"No Old Chestnuts"

Shortly after I was enrolled in the first grade of school, my older siblings declared that, following breakfast in our farmhouse, that particular sunny Saturday in October 1939 was scheduled for gathering the family's winter supply of wild nuts. It was a perfect morning for nutting—the grass white with the first heavy frost of autumn, a hint of the winter to come in the October-blue-sky morning. As each of us was verbally mapping out the sequence of visits to the best nut trees of the neighborhood—choice shagbark-hickories on our grandparents' farm, special black walnut trees with near baseball-sized globes, the best butternut trees, thickets of hazelnut bushes, big shellbarks with their king-sized nuts—Mother reminded everyone to be sure to include the fine old chestnut tree in Everett Water's pasture, the last known surviving American chestnut tree in our Ripley County neighborhood.

Since I had missed the previous two days of school with an early-season sore throat, Mother decided at the last moment that I was not permitted to join the nut-gathering excursion. Thus disappointed, I stayed home and watched through the dining room window, as off the older children went, with pails and burlap bags in hand, talking excitedly as they walked away.

As fate would have it, the next summer (of 1940), Mr. Water's chestnut joined in death the hundreds of millions of American chestnut trees throughout the forests of eastern North America that had succumbed to the exotic fungal pathogen *Cryphonectria parasitica*, known since as "the chestnut blight." So, thereby, I missed by one year the opportunity to partake of a great American tradition—the gathering of tasty wild American chestnuts.

Believed to have arrived about 1905 on infected nursery stock from Asia, the chestnut blight spread with alarming speed throughout most of the natural range of the American chestnut. By the early 1940s the vast majority of adult trees were dead, their ghostly silvery-gray trunks of extremely durable wood punctuating the slopes and ridgelines from Georgia to New England, and from the Atlantic to Indiana and beyond.

The parasite kills infected adult trees by forming cankers or lesions, which destroy the cambium of the lower trunk, effectively girdling and thereby killing the aboveground stem. The root system, however, may remain alive for years; oftentimes repeatedly producing viable root sprouts in an attempt to survive the infection. Usually these regrowths also become infected in due time, and die before reaching reproductive age. Such stump sprouts occasionally are still encountered in the "knobs" of southern Indiana, but rarely do they grow large enough to flower.

Today, it is quite a find to encounter a surviving wild adult tree that is healthy and bearing viable nuts. For all intents and purposes, the tree is gone from the American scene as a forest species. This is the one native tree species that has essentially been extirpated from Indiana as reproducing individuals. What a loss.

But the American chestnut may not be a doomed species in the long run. A few widely scattered, isolated trees were spared the infection. Other trees apparently were naturally resistant to the disease. Forest geneticists are now working diligently to develop blight-resistant forms from these genetic reservoirs, and have met with some success. Seedlings have been developed that have considerable promise in being able to survive into maturity. In fact, students in our Indiana State University Honors Seminar in Interdisciplinary Environmentalism helped Dr. Ed

Warner and me plant a fine young blight-resistant American chestnut tree on the ISU campus in April 2000. So far it looks very healthy, and may grow into an adult tree, giving us hope that the chestnut could be restored to our forests.

Perhaps we will, in time, come full circle in that magnificent young American chestnut trees will again lift their lofty crowns within the forestlands of southern Indiana, and throughout their natural range in the Eastern Deciduous Forest. Then, in some future autumn, another six-year-old Indiana boy can take a pail in hand and join his siblings in picking up American chestnuts on a frosty October morning, being careful all the while not to mistakenly prick his fingers on the spiny burs.

101 TREES OF INDIANA

Species List of Indiana Trees

Species are arranged here phylogenetically by family, genus, and species, which is the order in which they appear in the two sections of the book which list species descriptions. By listing species phylogenetically, as is typically done in most taxonomic manuals, closely similar species are grouped together, which facilitates species comparisons. Species codes and the nomenclature source are given at the end of this list. (All taxa considered herein are included in this list; however, some taxa have been extirpated, or are listed as excluded for various reasons.)

	Scientific Name	Common Name(s)	Family	Page(s)
E	*Ginkgo biloba* L.	Ginkgo; Maidenhair tree	Ginkgoaceae	288–289
E	*Picea abies* (L.) Karsten.	Norway spruce	Pinaceae	290
I	*P. pungens* Engelm.	Colorado blue spruce; Blue spruce	Pinaceae	291
	Tsuga canadensis (L.) Carrière.	Eastern hemlock	Pinaceae	82–83
	Larix laricina (Du Roi) K. Koch.	Tamarack; Eastern larch	Pinaceae	84–85
	Pinus strobus L.	Eastern white pine	Pinaceae	86–87
I	*P. resinosa* Aiton.	Red pine; Norway pine	Pinaceae	292
E	*P. nigra* J. Arnold.	Austrian pine	Pinaceae	295
E	*P. sylvestris* L.	Scotch pine; Scot's pine	Pinaceae	293

(continued)

Species List of Indiana Trees *(continued)*

	Scientific Name	Common Name(s)	Family	Page(s)
	P. banksiana Lambert.	Jack-pine	Pinaceae	88–89
	P. virginiana Miller.	Virginia-pine; Scrub-pine	Pinaceae	90–91
I	*P. echinata* Miller.	Shortleaf pine; Yellow pine	Pinaceae	294
	Taxodium distichum (L.) Rich.	Bald cypress	Taxodiaceae	92–93
	Thuja occidentalis L.	Northern white cedar; Arborvitae	Cupressaceae	94–95
	Juniperus virginiana L.	Eastern red cedar	Cupressaceae	96–97
	Magnolia acuminata (L.) L.	Cucumber-tree	Magnoliaceae	98–99
	M. tripetala (L.) L.	Umbrella-tree	Magnoliaceae	100–101
	Liriodendron tulipifera L.	Tulip-tree; Yellow poplar	Magnoliaceae	102–103
	Asimina triloba (L.) Dunal.	Pawpaw	Annonaceae	104–105
	Sassafras albidum (Nutt.) Nees.	Sassafras	Lauraceae	106–107
	Platanus occidentalis L.	American sycamore	Platanaceae	108–109
E	*P. orientalis* L.	Oriental planetree	Platanaceae	296
	Liquidambar styraciflua L.	Sweet gum; Red gum	Hamamelidaceae	110–111
	Hamamelis virginiana L.	Witch-hazel	Hamamelidaceae	297
	Ulmus americana L.	American elm; White elm	Ulmaceae	112–113
	U. rubra Muhl.	Slippery elm; Red elm	Ulmaceae	114–115
	U. thomasii Sarg.	Rock-elm; Cork elm	Ulmaceae	116–117

(continued)

3

Species List of Indiana Trees *(continued)*

(continued)

Species List of Indiana Trees (*continued*)

7

Species List of Indiana Trees (continued)

	Scientific Name	Common Name(s)	Family	Page(s)
E	*Acer platanoides* L.	Norway-maple	Aceraceae	317
	A. saccharum Marshall.	Sugar-maple	Aceraceae	260–261
	A. nigrum Michx. f.	Black maple	Aceraceae	262–263
	A. rubrum L.	Red maple	Aceraceae	264–265
	A. saccharinum L.	Silver-maple	Aceraceae	266–267
	Acer negundo L.	Boxelder; Ash-leaved maple	Aceraceae	268–269
S	*Rhus glabra* L.	Smooth sumac	Anacardiaceae	318–319
	R. typhina L.	Staghorn-sumac	Anacardiaceae	270–271
S	*R. copallinum* L.	Shining sumac	Anacardiaceae	320–321
S	*Toxicodendron vernix* (L.) Kuntze.	Poison-sumac; Swamp sumac	Anacardiaceae	322–323
E	*Ailanthus altissima* (Mill.) Swingle.	Tree of heaven; Ailanthus tree	Simaroubaceae	324–325
S	*Zanthoxylum americanum* Miller.	Common prickly ash; Toothache tree	Rutaceae	326–327
S	*Ptelea trifoliata* L.	Common hop-tree; Wafer-ash	Rutaceae	328
S	*Aralia spinosa* L.	Hercules' club; Devil's-walkingstick	Araliaceae	329
S	*Forestiera acuminata* (Michx.) Poiret.	Swamp-privet	Oleaceae	330
	Fraxinus americana L.	White ash	Oleaceae	272–273
Exc.	*F. biltmoreana* Beadle.	Biltmore ash	Oleaceae	339
	F. pennsylvanica Marshall.	Green ash (Red ash)	Oleaceae	274–275

Code	Species	Common name	Family	Pages
Exc.	*F. lanceolata* Borckhausen.	(Green ash)	Oleaceae	339
	F. profunda (Bush) Bush.	Pumpkin-ash	Oleaceae	276–277
	F. nigra Marshall.	Black ash	Oleaceae	278–279
	F. quadrangulata Michx.	Blue ash	Oleaceae	280–281
E	*Paulownia tomentosa* (Thunb.) Steudel.	Empress-tree; Royal paulownia; Princess tree	Bignoniaceae	331
	Catalpa speciosa Warder.	Northern catalpa	Bignoniaceae	282–283
I	*C. bignonioides* Walter.	Southern catalpa	Bignoniaceae	332
S	*Viburnum lentago* L.	Nannyberry; Sheepberry	Caprifoliaceae	333
S	*V. rufidulum* Raf.	Southern black haw	Caprifoliaceae	334–335
S	*V. prunifolium* L.	Black haw	Caprifoliaceae	336

Species codes as follows:

I—Introduced into Indiana, but native to North America.
E—Exotic species growing in Indiana.
Ex—Extirpated Species.
Exc.—Excluded Species, Extirpated or Subject to Taxonomic Revision.
S—Shrub-sized typically; occasionally reach tree size.

Total Tree Flora:
34 Families
62 Genera
144 Species

Native Tree Flora:
101 Tree Size Species
13 Shrub-Tree Species

Source: Nomenclature follows the *Manual of Vascular Plants of Northeastern United States and Adjacent Canada*, 2nd edition, by Henry A. Gleason and Arthur Cronquist, published in 1991 by The New York Botanical Garden (Bronx, N.Y. 10458): 910 p.

Introduction: Purpose and Plan of the Book

This book was written in response to the need for a field manual for identifying Indiana's trees. Both the popular booklet *Fifty Trees of Indiana* by T. E. Shaw, and the more technical third edition of *Trees of Indiana* by Charles C. Deam and "Ted" Shaw (published in 1953), were out of print and largely unavailable for several years, although both have been re-issued recently, without their texts being updated.

Our aim was to produce a book that covers essentially all of Indiana's tree species, yet is usable by people with a wide range of outdoor interests—from high school students identifying trees for their science projects, to individual landowners, to professional naturalists inventorying natural areas, to professors and college students in taxonomy or dendrology courses. Those who use the manual over time will determine how successful we were with this endeavor.

No attempt was made to include all the excellent botanical detail of Deam's text, nor do we include all the varieties, forms, and hybrids that he recognized. Not that such information lacks importance; it is beyond the scope of this book, and best left to consideration by professional botanists.

Two questions fundamental to a work such as this are: What is a tree? and Which tree species should be included? These have not been answered totally; perhaps they are irresolvable.

We define a tree as a woody perennial that *typically* has a single trunk, and potentially grows to a diameter greater than 4 inches (as measured at 4 1/2 feet above the ground), and/or reaches 20 or more feet tall at maturity. These criteria allow us to include a number of large shrubs that may, on occasion, grow to tree size. Examples are Speckled alder, Hercules' club, Shining and Smooth sumac, Alternate-leafed dogwood, Wahoo, Swamp-privet, Autumn olive, buckthorns, Wafer-ash, the black haws, and Common prickly ash.

Which trees to include (or more accurately to exclude) was more vexing. Essentially all tree species native to Indiana, and with populations still viable in the state, were

automatically included. Also included are several introduced species such as Norway-maple, Osage-orange, White
mulberry, common catalpa, Tree of heaven, Princess tree,
and "Mimosa," which are widespread in certain areas,
reproducing on their own, and apparently naturalized.
Other commonly planted and frequently encountered
species such as Red pine, Scotch pine, Norway spruce,
Siberian elm, Horse-chestnut, Weeping willow, and others
are described in less detail.

The distribution maps, which accompany each species
description, indicate the counties in which that tree has
been found and recorded. These maps have been updated to
include more than 2000 new county records discovered by
scientists, foresters, and naturalists since *Trees of Indiana*
was published in 1953, and by recent systematic field
inventories of all counties by the author of this volume.
These new records are based on positive identifications, but
many are not verified by herbarium specimens. Since
information of this type is always being refined, diligent
field searches will produce additional county records for
many species. Distribution changes for invading species,
especially, will create others.

The species distribution maps are keyed as follows: (1)
the solid round dots are the original county records listed
by Deam and Shaw in their 1953 *Trees of Indiana;* and (2)
the solid triangles are county records discovered during the
past 50 years.

Whenever possible, sketches are originals drawn by
Katherine Harrington from live or collected materials. In
some cases, previous illustrations were re-drawn and
adapted to our specific needs for this book.

Photographs are included to show field characteristics
for each genus, and most individual tree species. In addition
to their obvious identification value, they give users of the
manual a look at the aesthetic beauty and form of Indiana's
trees. The majority of the photographs are outstanding
images from the vast dendrological color slide library of
Ron Rathfon, Extension Forester, Purdue University.

We elected not to follow the metric system of measure-
ment for the reason that most potential users of this manual
will be more familiar with the English system. Conversion
scales and tables for comparing the two systems of mea-
surement are conveniently located inside the back cover.

Two pages are typically devoted to each species, both of

which, as facing pages, are visible at one time. The sketches show greater and more precise year-round detail than is ever possible from a single photograph taken in one season. Multiple photographs of different features permit verification of species identified by keys, descriptions, or sketches. Descriptions of species characteristics attempt to provide a range of material useful in identification, yet in wording simple enough not to intimidate the beginning student of trees. Moreover, verbal characteristics usable in all seasons are included. Identification keys were kept short enough not to entail a waste of time following blind alleys, yet inclusive enough to be useful.

This book was compiled with the intention that it be used—both widely and often. Hence its design and shape. Its long, narrow format adapts it to convenient field use; it slips easily into a hip pocket, purse, field guide case, or daypack. It feels good in the hand—like an *Estwing* claw hammer. A tool to be used, and enjoyed.

INDIANA: AN OVERVIEW

The Indiana Landscape

Indiana lies near the center of the Eastern Deciduous Forest Biome, which encompasses approximately the southeastern quarter of the North American Continent. At 36,291 square miles, Indiana is slightly smaller in size than the average for states located east of the Mississippi River.

It ranges north–south from the Ohio River at 37°40' to the state of Michigan at 41°50' north latitude, for a length of about 275 miles. In longitude it lies between 84°49' on the east and 88°2' on the west, for a maximum width of nearly 180 miles. Elevation ranges from the highest point in the state in Randolph County at 1257 feet above mean sea level, to the mouth of the Wabash River in Posey County at 313 feet. The average elevation is about 700 feet (Deam, 1953).

Except for the more rugged hill country of south-central Indiana, the majority of the state has been glaciated, much of it by multiple ice advances. Extreme northern Indiana was deglaciated most recently, likely about 13,000 years ago (Melhorn, 1997). In general, the topography has less relief in glaciated regions than is the case for areas that escaped the impact of the ice.

The average annual precipitation is about 40 inches, with total annual snowfall ranging from 60 inches near Lake Michigan to less than 10 inches at Evansville. The average annual temperature is about 52 degrees Fahrenheit; the growth season varies from an average of less than 160 days in the northeastern corner to about 200 days at the extreme southwestern tip (Newman, 1997).

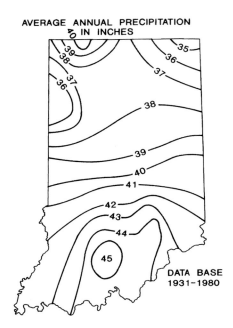

AVERAGE ANNUAL PRECIPITATION
IN INCHES

DATA BASE
1931-1980

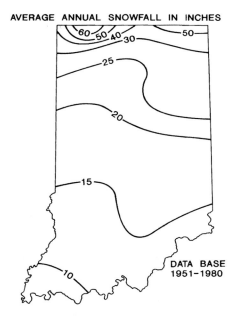

AVERAGE ANNUAL SNOWFALL IN INCHES

DATA BASE
1951-1980

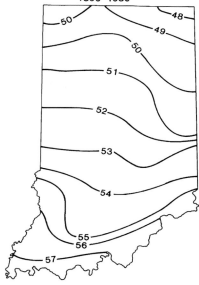

ANNUAL AVERAGE TEMPERATURE °F
1890–1980

48
50
49
50
51
52
53
54
55
56
57

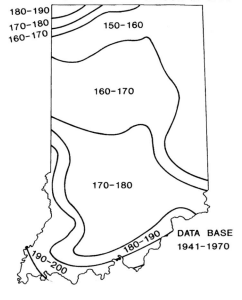

AVERAGE ANNUAL GROWING SEASON IN DAYS

180–190
170–180
160–170
150–160
160–170
170–180
180–190
190–200

DATA BASE
1941–1970

The state has been divided into 12 natural regions, which reflect differences in bedrock geology, glacial history, soils, climate, vegetation, and vertebrate fauna. Homoya et al. (1985) defined a natural region as a "generalized unit of the landscape where a distinctive assemblage of natural features is present." Each of the natural regions is described more fully in Jackson (ed.), 1997. For a brief discussion of how individual tree species respond to environmental differences, see, in this volume, pages 24–28.

THE NATURAL REGIONS OF INDIANA

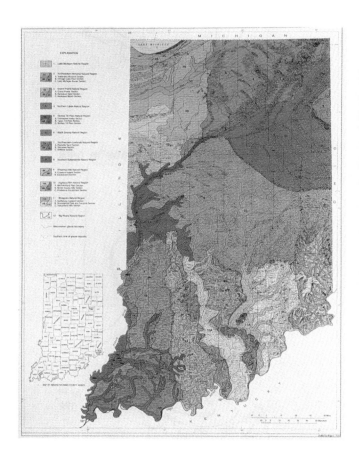

Indiana has 92 counties, ranging in size from Allen County at 657 square miles to Ohio County at 87 square miles. This county outline map serves as a locator for individual tree distribution maps for each tree species considered herein.

COUNTY MAP OF INDIANA

Indiana Forestlands and Tree Flora

During presettlement years (prior to 1800), of the 23,227,000 acres which now constitute Indiana, about 20 million acres (87%) were originally forested. The remaining portion was prairie, wetlands, water, and other non-forested vegetation types (Petty and Jackson, 1966).

About half of the state was covered by forests dominated by American beech and Sugar-maple; nearly 30% by some combination of several oak and hickory species; with more than 7% of a more mixed forest of Appalachian origin (western mesophytic) (Lindsey et al., 1965).

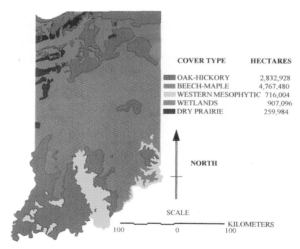

COVER TYPE	HECTARES
OAK-HICKORY	2,832,928
BEECH-MAPLE	4,767,480
WESTERN MESOPHYTIC	716,004
WETLANDS	907,096
DRY PRAIRIE	259,984

NORTH

SCALE

KILOMETERS
100   0   100

Presettlement Vegetation Map of Indiana (after Lindsey et al., 1965).

In the words of Bramble and Miller (1966),

> As seen by early French explorers, the great forests
> which sprawled from the lower tip of Lake Michigan, far
> southward to the Ohio River, were a magnificent yet
> forbidding wilderness of giant hardwoods. . . . Forty-two
> species of hardwoods in the Wabash Valley were reported to
> reach a height of one hundred fifty feet or more. . . . With
> the exception of local concentrations of coniferous trees,
> plus the more extensive prairies and sloughs of the north,
> scattered meadows and Indian clearings along the Wabash,
> the great hardwoods dominated and overwhelmed as far as
> the eye could reach. . . . As isolated settlements sprang up,
> the great trees began to fall. Hundreds were felled, rolled
> into the closest ravine and burned, to make room for the
> plow.

By 1917, the state's forested land had shrunk to
1,660,000 acres (slightly more than 7% of the total area),

causing State Forester Charles C. Deam to predict in 1922 that the state would be treeless in 15 years. Fortunately, Deam was wrong at least once! Today, about 4.8 million acres of Indiana are forested, or a little more than 20%. Since the 1960s, Indiana's forests have shown a yearly net increase in acreage, an encouraging trend that hopefully will continue. Presently, private landowners own about three-quarters of the forestland in the state (in some 100,000 tracts, most of them relatively small), with the remaining quarter about evenly divided among federal, state, and corporate ownership.

During Deam's tenure as state forester, the General Assembly of Indiana passed the Forest Classification Act in 1921, which offered a very low real estate tax for landowners who would, under a written agreement with the state forester, enter into a basic program of forest protection and management. Long known as Classified Forests, such designated lands now contain some of the best-forested tracts in the state. Today more than 13,000 separate tracts totaling nearly 16% of Indiana's forestlands (more than 760,000 acres) are designated as Indiana Classified Forests.

The vast majority (96%) of Indiana's present-day forestland is hardwoods, with oak-hickory (at 32%)–and beech-maple (at 21%)–dominated stands the major types. A variety of other compositions make up the balance.

In addition, trees obviously occur in an array of non-forested landscape types including fencerows, stream borders, abandoned fields, wetlands, farmlands, parklands, and urban-residential. Most anywhere your outdoor rambles take you in Indiana, you will not be far from trees.

Of the 101 species of native Indiana trees that routinely get tree-sized, one-fourth (25 species) are either oaks or hickories of the genera *Quercus* and *Carya*. Overall, of the 54 genera of native species that, at least on occasion, reach tree size in Indiana, more than 60% of the species belong to only 11 genera, namely: *Quercus, Carya, Acer, Fraxinus, Ulmus, Prunus, Populus, Crataegus, Salix, Betula,* and *Pinus.* Once the characteristics of the 8 genera, each having 4 or more species, are mastered, you will have a greater than 50% chance of being able to identify any native tree species that you encounter.

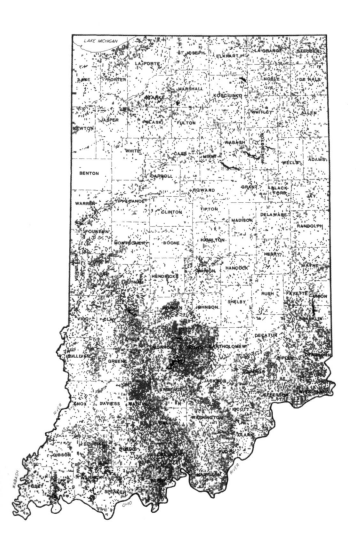

Indiana's Present Forested Areas

Ecological Relationships among Trees

The presence or absence of trees (as is the case for all organisms) at a given location is based on the occurrence and availability of the necessary resources of energy, nutrients, water, and space, and on the individual species' ability to compete for those resources. Each species has somewhat different requirements for resources; hence tree species are able to co-exist, and to assemble into a forest community at a given site. That each species requires and utilizes resources differently is a basic premise of niche theory, one of the fundamental ideas of ecology.

The availability of resources obviously changes over time, resulting in a succession of tree communities that occur sequentially at a given location on the landscape. For example, disturbance by natural events or human-induced changes has greatly altered or eliminated the vast majority of the original forest that greeted the first settlers in Indiana. As old-growth forests were cleared for farming and development purposes, open space was created in which pioneer trees could invade and begin rebuilding the forests that had been removed. The process of clearing allowed abundant sunlight energy to reach ground level, thereby creating very different conditions for forest regrowth than had existed under the dense canopies of the original trees.

The process of rebuilding ecological communities following their complete removal or extensive disturbance is called ecological succession. Although this is a gradual and continual series of changes (continuum), for convenience these recovery communities can be divided into pioneering, mid-successional, and near-equilibrium or climax stages. Similarly we have a continuum of environmental conditions represented in Indiana from open water, to wetlands, to moist slopes and uplands, to dry ridges, to very dry dune sands, rocky slopes, and cliffs. Again for convenience, we can group the moisture gradient into wet (hydric), moist (mesic), and dry (xeric) sites. The resultant three-by-three habitat categorization, according to the time and moisture gradients, results in nine basic environmental situations in which trees can grow; for example, xeric pioneer, mesic mid-successional, hydric climax, or any of the other six possible combinations. Viewed in this manner, patterns quickly emerge as to which habitat should be searched to locate a given set of tree species or, viewed ecologically, to find a given forest community type.

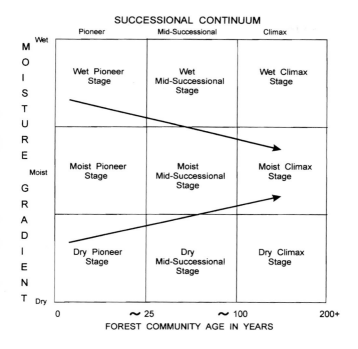

Arrows indicate the tendency for succession patterns to converge toward a Moist (Mesic) Climax Community as landscapes mature and environmental conditions become more intermediate over time.

The availability of light at the seedling level, plus the relative abundance of soil moisture and soil nutrients, are the primary determinants of which tree species are able to gain the advantage within a given environment. For example, abandoned farm fields and fencerow habitats typically are invaded quickly by such shade-intolerant to intermediate tolerant species as tulip-tree, sycamore, elm, sassafras, wild black cherry, boxelder, and eastern red cedar, all of which are efficiently dispersed to those sites, either by wind or by birds, and all require high-light environments for seedling survival. Mid-successional species such as the ashes, oaks, hickories, walnut, and white pine require medium light conditions as they assemble into forest communities on sites of intermediate age. These species do not reproduce effectively in the shaded environment created at the forest floor by the large canopy trees. Old-growth forests, which often are dominated in Indiana by American beech and sugar maple, typically terminate forest successions on moist slopes and well-drained uplands, frequently with tulip-tree, basswood, red elm, and ash as sub-dominant species. Seedlings of these dominant species are shade tolerant, which permits them to survive within the low-light conditions created by the dense canopy of the adult trees, thereby perpetuating the canopy species at that site. Such mature communities, which represent the endpoint of the successional continuum, are often called equilibrium or climax communities, although they never reach complete stability.

By searching habitats that represent the range of moisture conditions from wet to moist to dry; and by examining successional stages representing early, mid-successional, and climax forest communities, most of the major tree species that are widespread within the state will be encountered. But to find those species with highly restricted geographic ranges (e.g., yellow-wood, northern white cedar, or water-locust) requires visits to the specific sites where they occur naturally.

Ecological Relationships among Selected Trees: Site-Moisture Adaptation Categories

Very Wet
Bald cypress
Silver-maple
Black ash
Pumpkin-ash
Water-locust
Tamarack
Cottonwood
Swamp-cottonwood
Swamp white oak
Overcup-oak
Black willow

Wet
Boxelder
Red maple*
Ohio-buckeye
River birch
Pecan
Shellbark-hickory
Sugarberry
Northern hackberry
Green ash
Honey-locust
Sweet gum
Black gum
American sycamore
Bur-oak*
Swamp chestnut-oak
Pin-oak
Shumard oak*
American elm
Post-oak*

Moist
Sugar-maple
Pawpaw
Yellow birch
Blue beech
Bitternut-hickory
Redbud
American beech
White ash
Black walnut
Kentucky coffee-tree
Butternut
Tulip-tree
Cucumber-tree
Northern red oak
Basswood
Eastern hemlock
Slippery elm

Dry
Red maple*
Downy serviceberry
Paper birch
Pignut-hickory
Shagbark-hickory
Mockernut-hickory
Northern catalpa
Flowering dogwood
Persimmon
Blue ash
Hop-hornbeam
Eastern white pine
Big-toothed aspen
Quaking aspen
Wild black cherry
Sweet crabapple
Shumard oak*
White oak
Shingle-oak
Chinkapin-oak
Black locust
Sassafras
Bur-oak*

Very Dry
Eastern red cedar
Jack-pine
Virginia-pine
Scarlet oak
Northern pin-oak
Southern red oak
Black-jack oak
Rock chestnut-oak
Post-oak*
Black oak
Winged elm
Rock-elm

*Note: Bimodal species.

27

Shade Tolerance of Selected Indiana Tree Species

Conifers

Very Tolerant:
>Eastern hemlock

Tolerant:
>Northern white cedar
>Norway spruce

Intermediate:
>Eastern white pine
>Bald cypress

Intolerant:
>Eastern red cedar
>Red pine
>Shortleaf pine
>Virginia-pine

Very Intolerant:
>Tamarack
>Jack-pine

Hardwoods

Very Tolerant:
>Hop-hornbeam
>Hornbeam
>American beech
>Sugar-maple
>Flowering dogwood

Tolerant:
>Red maple
>Silver-maple
>Boxelder*
>Basswood
>Black gum
>Persimmon*
>Buckeyes

Intermediate:
>Yellow birch
>American chestnut
>White oak
>Northern red oak
>Black oak
>American elm
>Slippery elm
>Rock-elm
>Northern hackberry
>Magnolias*
>White ash
>Green ash
>Black ash

Intolerant:
>Black walnut
>Butternut
>Pecan
>Hickories
>Paper birch
>Tulip-tree
>Sassafras
>Sweet gum
>American sycamore
>River birch
>Wild black cherry
>Honey-locust
>Kentucky coffee-tree
>Catalpas

Very Intolerant:
>Willows (as a genus)
>Quaking aspen
>Big-toothed aspen
>Cottonwood
>Gray birch
>Black locust
>Osage-orange

Source: Adapted from Baker, *Principles of Silviculture* (1950).
Note: Species of uncertain status marked with an asterisk (*).

NAMING AND IDENTIFYING TREES

How Trees Are Named

Every kind of tree found in Indiana has one or more common names and a single scientific or Latin name. Common names have the advantage of being generally known to laypersons, are typically easily pronounced, spelled, and remembered, and may be descriptive of one or more characteristics of that particular type of tree; for example, White oak (named for its light-colored wood), Paper birch (for its peeling, papery, white bark), or Tulip-tree (for the shape of its leaves and flowers). Disadvantages of common names include: they vary from one locality to another for a given kind of tree, some trees have more than one common name, and the same common name may be given to quite dissimilar trees.

In contrast, scientific names for trees, usually written in Latin (a dead and unchanging language), are universally recognized, accepted, and used by scientists and foresters throughout the world. Furthermore, they are frequently even more descriptive of a given tree's characteristics than are common names. For example, the scientific name of White oak is *Quercus alba* from the Latin for oak and white, respectively; that of Paper birch is *Betula papyrifera*, from the Latin for birch and paper-like, respectively; and that of Tulip-tree is *Liriodendron tulipifera*, literally meaning a Tulip-tree (*Lirio*, "tulip," and *dendron*, "tree"), which bears "tulips" (*tulipifera*).

The order of scientific names is reversed from that used for common names. For *Quercus alba*, *Quercus* corresponds to a person's surname (e.g., Deam), while *alba* corresponds to the individual's given name (e.g., Charles). Hence, "Deam, Charles" instead of "Charles Deam," the latter following the order of common names for trees. It follows then that *Quercus* (the generic name) is a collective term for all of the oak species combined; whereas *alba* (the species name) identifies one individual type of oak.

The third element of a scientific name is an abbreviation of the name of the botanist who first named that species of tree; for example, for *Quercus alba* L., the L. stands for Carolus Linnaeus, the Swedish botanist who first named the tree during the eighteenth century. Likewise, Michx. refers to André Michaux, a nineteenth-century botanist.

Trees have relatives, just as people do. All species of oaks found in Indiana belong to the same genus (*Quercus*) and may be thought of as being first cousins. Although they have different parents, they have strong genetic similarities. At the next higher classification level, all of the oaks belong to the same family (Fagaceae or beech family) as does the American beech (genus *Fagus*) and the American chestnut (genus *Castanea*). Hence, we can consider the oaks to be second cousins to the beeches and chestnuts. All Indiana tree species are classified according to their relatedness. Because of these relationships, trees, typically, are first recognized to family, then to genus, and finally to species. These classification units and their recognition are the basis for identification as described in the next section.

How Trees Are Identified

Each kind (species) of tree that occurs in Indiana is different from every other. Tree species differ in appearance because they differ genetically—each has DNA that is unique. Differences in DNA result in differences in leaves, bark, twigs, buds, flowers, fruit, and seeds—all of which are characteristics easily recognized and evaluated by persons interested in identifying trees. Just as you recognize your friends and family members by how they differ in size, body shape, hair color, and the like, so can you also learn to recognize different tree species. Since many trees are quite similar and have only subtle differences, you must learn to observe them very closely, and know what to look for.

Leaves: Leaves differ primarily in their arrangement on the stems, whether their structure is simple or compound, and in their overall shape. Other leaf characteristics involve whether their edges (margins) are smooth, toothed, or lobed, the length of their stems (petioles), the shape of leaf tips or bases, or surface characteristics. Representative illustrations follow:

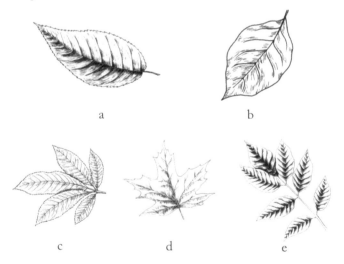

a   b

c   d   e

Leaf types:
(a) simple-serrate;
(b) simple-smooth;
(c) compound-palmate; (d) simple-palmate;
(e) compound-pinnate

Bark: Although the bark of individual trees varies greatly from young to old age, from one ecological site to another, or under different growth conditions, with practice it is quite easy to identify individual trees accurately by quickly glancing at bark characteristics. Experienced foresters routinely recognize trees, even at a distance, by noting bark features of the trees they encounter. Barks of individual tree species differ in color, texture, thickness, width and pattern of furrows and ridges, scaly or lack of scaliness, hardness, and so on.

Crowns: Likewise, many species have distinctive branching patterns and overall sizes and shapes that enable an experienced outdoorsperson to recognize or identify trees by silhouette, even at a distance.

Twigs/Buds: Especially during leafless periods, twig and bud characteristics become extremely valuable for accurate identification of many tree species. Arrangement of leaf scars on a twig in winter reflects leaf arrangement in summer. Similarly, the position, shape, size, color, and hairiness of buds, plus the number and arrangement of scales on the buds of a given tree species are like those found on no other species. Vascular bundle scars and pith characteristics also are often important. Some examples follow:

Twig/Bud types:
(a) alternate-scaled;
(b) terminal-clustered;
(c) alternate-smooth;
(d) opposite-scaled

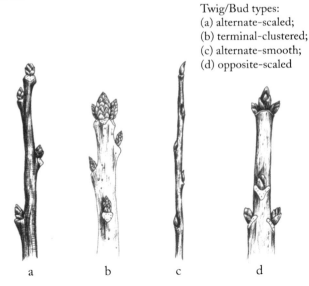

a   b   c   d

Flowers/Fruit/Seeds: Plant reproductive structures have long been used by botanists for classifying species because these structures typically vary less than do such vegetative structures as leaves, bark, twigs, and buds. Detailed examination of the arrangement and structure of individual flower parts usually is not necessary for field identification of most tree species; the general arrangement, overall structure, and determination of whether flowers are unisexual or bisexual is usually sufficient.

Tree fruits are of two basic types, dry or fleshy. Dry fruits include such nuts as walnuts, hickory nuts, or the acorns of oaks; dry winged fruits of maple, ash, or elm; or the "pods" of Black locust, Kentucky coffee-tree, or catalpas. Examples of fleshy fruits are those of Wild black cherry, Persimmon, mulberries, and hawthorns. Examination of fruit and seed structure is extremely useful in identifying many species.

Use of Keys

Identification keys are organized in such a way that tree characteristics may be compared in a systematic manner. By making choices in a logical sequence, the identity of a tree can be narrowed down first to family or genus, then determined to the individual species. Keying out tree species is similar to the way two friends might play "20 Questions" in order to determine which of their classmates was seen at the shopping mall.

More specifically, two general keys (for summer and winter use) are presented to assist the student of trees in identifying an unknown tree to genus. The "Key to Genera" is arranged with pairs of leads. Each lead describes contrasting characteristics.

To determine the identity of a tree, read the first lead. If that lead fits your specimen, proceed to succeeding pairs of leads in sequence until the name of a genus is reached. If the first lead does not fit your tree, then try the contrasting lead and accept or reject these leads sequentially until the key leads you to a genus.

For **those 12 genera that contain three or more species each,** abbreviated keys to species (in both summer and winter condition) are included for each such genus. When you have identified a specimen to the species level, confirm your determination by studying the illustrations, descriptions, and photographs in the species descriptions in either section 1 or section 2 of the book text. Species within genera having only two species each may be determined without the use of a separate key to species by comparing the illustrations and descriptions for each.

The task of keying specimens may not be as easy as it first seems, especially for closely similar genera or species. After you have followed the key to a genus or species, you may find that the tree selected does not fit the characteristics of that taxon. Then you must retrace the steps taken and be more careful, being particularly sure that the terms used in the key are understood by consulting the glossary beginning on page 353. Errors in using keys are usually the result of making prejudgments as to the specimen's identity, of misunderstanding the terms used, of having poor or inadequate material for naming, or due to haste or inexperience. Rely heavily on the illustrations and descriptions as aids to identification by keying, but not as a replacement for the keys.

In this manual to the trees of Indiana, inclusive keys to all species considered herein are identifiable to the generic level by following either "Division 1: Trees with Leaves Present" (basically a Summer Key) or "Division 2: Trees with Leaves Absent" (basically a Winter Key).

Identification Keys

Keys to Genera of Indiana Trees

Division 1: Trees with Leaves Present

GROUP 1. Trees with needle-like, scale-like, or awl-like leaves

    Juniperus, Larix, Picea, Pinus, Taxodium, Thuja, Tsuga

GROUP 2. Trees with opposite, compound broad leaves

    Acer (*A. negundo,* only), Aesculus, Fraxinus

GROUP 3. Trees with opposite, simple broad leaves

    Acer, Catalpa, Cornus (*C. florida,* only), Euonymus, Forestiera, Paulownia, Rhamnus (*R. cathartica,* only), Viburnum

GROUP 4. Trees with alternate, compound broad leaves

    Ailanthus, Aralia, Carya, Cladrastis, Gleditsia, Gymnocladus, Juglans, Koelreuteria, Ptelea, Rhus, Robinia, Toxicodendron, Zanthoxylum

GROUP 5. Trees with alternate, simple broad leaves

    Alnus, Amelanchier, Asimina, Betula, Carpinus, Castanea, Celtis, Cercis, Cornus (*C. alternifolia,* only), Crataegus, Diospyros, Elaeagnus, Fagus, Ginkgo, Hamamelis, Liquidambar, Liriodendron, Maclura, Magnolia, Morus, Nyssa, Ostrya, Oxydendrum, Platanus, Populus, Prunus, Pyrus, Quercus, Rhamnus (*R. caroliniana,* only), Salix, Sassafras, Tilia, Ulmus

Division 2: Trees with Leaves Absent

GROUP 1. Not applicable for these genera, except for Larix and Taxodium, which are winter deciduous

GROUP 2. Trees with opposite or whorled leaf scars on twigs

    Acer, Aesculus, Catalpa, Cornus (*C. florida,* only), Euonymus, Forestiera, Fraxinus, Ginkgo, Larix, Paulownia, Rhamnus (*R. cathartica,* only), Viburnum

GROUP 3. Trees with alternate leaf scars on twigs

    Ailanthus, Albizia, Alnus, Amelanchier, Aralia, Asimina, Betula, Carpinus, Carya, Castanea, Celtis, Cercis, Cladrastis, Cornus (*C. alternifolia,* only), Crataegus, Diospyros, Elaeagnus, Fagus, Gleditsia, Gymnocladus, Hamamelis, Juglans, Koelreuteria, Liquidambar, Liriodendron, Maclura, Magnolia, Morus, Nyssa, Ostrya,

Oxydendrum, Platanus, Populus, Prunus, Ptelea, Pyrus, Quercus, Rhamnus (*R. caroliniana,* only), Rhus, Robinia, Salix, Sassafras, Taxodium, Tilia, Toxicodendron, Ulmus, Zanthoxylum

Key to Genera

Division 1: Trees with Leaves Present

1. Leaves needle-like or scale-like ("evergreen" trees) 2
   Leaves broad (deciduous trees) 8

2. Leaves scale-like (or some branches with awl-shaped leaves) 3
   Leaves needle-like (linear); cones woody 4

3. Leaves all scale-like; cone small, woody with overlapping scales .... *Thuja*
   Leaves scale-like, or some awl-shaped; cone fleshy, berry-like .... *Juniperus*

4. Leaves needle-like and solitary 5
   Leaves needle-like in clusters of 2 to more than 10 7

5. Needles blunt at tip, with white lines below, in two ranks .... *Tsuga*
   Needles sharp-pointed 6

6. Needles soft, flat in cross-section, 1/2" long or less .... *Taxodium*
   Needles stiff, very sharp, square in cross-section, >1/2" long .... *Picea*

7. Needles in clusters of >5; borne on spur shoots on branches .... *Larix*
   Needles in bundles of 2–5, wrapped at base of needles .... *Pinus*

8. Leaves opposite 9
   Leaves alternate 19

9. Leaves compound 10
   Leaves simple 12

10. Leaflets 5–7, arranged hand-like (palmately); flowers

in showy pyramidal clusters; fruit large, nut-like ("buck-eye") in fleshy capsule .... *Aesculus*
    Leaflets 3–11, arranged ladder-like (pinnately); flowers and fruit otherwise 11

11. Leaflets 3 or 5; fruit winged in pairs; key-shaped propellors (maple "keys") .... *Acer* (*A. negundo*, only)
    Leaflets 5–11; fruit winged, single, paddle-shaped .... *Fraxinus*

12. Leaf stems (petioles) > 1 1/2" long 13
    Leaf stems <1 1/2" long 15

13. Leaf blades palmately 3–5 lobed .... *Acer*
    Leaf blades heart-shaped, margins smooth (entire) .. 14

14. Leaves in whorls of three, smooth to softly hairy; fruit long, round .... *Catalpa*
    Leaves opposite, densely and velvety hairy; fruit a short, oval capsule .... *Paulownia*

15. Leaves smooth to slightly toothed, broadly oval to obovate 16
    Leaves definitely toothed, oval to elliptical 17

16. Leaves opposite or nearly so; twigs spine-tipped; fruits small, black .... *Rhamnus* (*R. cathartica*, only)
    Leaves always opposite; bracts of flowers showy white; fruits bright red .... *Cornus* (*C. florida*, only)

17. Twigs green with longitudinal lines; fruits red-purple, bursting when ripe .... *Euonymus*
    Twigs gray-brown; fruits purple to black; leaf margins finely toothed 18

18. Leaves lanceolate; leaf tips long and tapering; fruit purple (to 3/4"), one seed .... *Forestiera*
    Leaves oval, leaf tips short tapered; fruits dark blue to black (1/2–3/4") .... *Viburnum*

19. Leaves compound 20
    Leaves simple 33

20. Leaves typically twice compound (bi-pinnate) 21

21. Leaf petioles and plant stems with spines; leaves very large (to 36"); fruits are juicy black berries (1/4"), several seeded .... *Aralia*
　　Leaf petioles and plant stems without spines (but stems or trunks may have thorns); fruits are pods (legumes) 22

22. Leaves huge (to 3' long); twice pinnate, larger (to 2" long) leaflets often alternate; fruit a blunt, stout pod .... *Gymnocladus*
　　Leaves smaller; usually twice pinnate; leaflets small (<1" long), opposite; fruit a thin, flat pod 23

23. Leaves once- or twice-compound; leaflets round-tipped; flowers "pea-like," in clusters; branches and trunks often with thorns .... *Gleditsia*
　　Leaves always twice-compound; leaflets sharp-pointed; flowers fluffy, silk-like, often pink in centers .... *Albizia*

24. Leaflets 3, without stems (sessile); fruit circular, flat, dry, one-seeded, winged .... *Ptelea*
　　Leaflets 5–41, stemmed or not; fruit various 25

25. Leaflets arranged alternately on leaf stem; fruit a thin pod .... *Cladrastis*
　　Leaflets opposite (pinnate) on leaf stem 26

26. Leaflets notched or toothed 27
　　Leaflets entire or essentially so 30

27. Leaflets notched; golden panicle of flowers; fruit a nutlet within a papery husk .... *Koelreuteria*
　　Leaflets toothed only 28

28. Leaflets 11–25; fruits are clusters of dry dull-red berries .... *Rhus*
　　Leaflets 5–23; fruits are large nuts 29

29. Leaflets 5–17 (typically 5–9); husk of fruit opens at maturity .... *Carya*
　　Leaflets 7–23 (typically 11–19); husk of round or oblong green fruit does not open at maturity .... *Juglans*

30.　Leaflets 13–41, entire plant smelly; fruit winged, in large clusters; pith of twigs very large .... *Ailanthus*
　　Leaflets 5–17, plant usually not smelly; fruit a small white berry, a pod, or small capsule; pith smaller 31

31.　Leaflets 7–13; twigs stout, warty; berries (to 1/4") white or creamy; found in bogs **(This plant is extremely toxic—<u>Do Not Touch!</u>)** .... *Toxicodendron*
　　Leaflets 5–17; paired spines at leaf bases; fruit a pod or small capsule 32

32.　Leaflets 5–11, sessile; fruits are small reddish capsules .... *Zanthoxylum*
　　Leaflets 7–17 with short stems; fruits are thin slender pods .... *Robinia*

33.　Leaves edges smooth (entire) 34
　　Leaves toothed to deeply lobed 45

34.　Trees with thorns or thorny branches 35
　　Trees without thorns; fruit much smaller 36

35.　Twigs zigzag with short, sharp, axillary spines; sap milky; fruit large (to 6" diam.), spherical; leaves glossy-green .... *Maclura*
　　Twigs usually straight with sharp ends; sap not milky; fruit small, berry-like, with stone; leaves silvery below .... *Elaeagnus*

36.　Leaf stems (petioles) long (>2") 37
　　Leaf stems shorter (<2") 38

37.　Leaves heart-shaped (cordate); flowers pea-like, showy pink; fruit a legume ("pod") .... *Cercis*
　　Leaves fan-shaped, notched at top; fruit plum-like, smelly .... *Ginkgo*

38.　Leaf blades usually longer than 6" 39
　　Leaf blades usually less than 6" long 40

39.　Leaves to 1' long, singly along stem; flower before leafing; fruit a large (to 4"), pulpy berry .... *Asimina*
　　Leaves usually 6" to 9" long, singly or in umbrella-like

whorls; flower after leafing; fruit a dry cluster with scarlet seeds .... *Magnolia*

40.   Leaves usually obovate, crowded toward branch tips; fruit a dark blue drupe .... *Nyssa*
      Leaves usually ovate, spaced along branches; fruit various 41

41.   Leaves 3–6" long, strap-like or an elongated oval 42
      Leaves usually <3" long, more oval in outline 44

42.   Leaves leathery with a blunt tip; fruit an acorn .... *Quercus* (*Q. imbricaria*, only)
      Leaves thinner with tapered tip; fruit otherwise 43

43.   Leaves are broadly oval, with arcuate (candelabra-like) veins; flat-topped clusters of flowers; fruits blue-black drupes .... *Cornus* (*C. alternifolia*, only)
      Leaves typically oval, not with arcuate veins; flowers and fruit otherwise 44

44.   Leaves are narrow ovals; flower clusters showy white, spreading, with urn-shaped florets; fruits are tiny capsules .... *Oxydendrum*
      Leaves typically oval; flowers inconspicuous; fruit an orange pulpy berry (persimmon) .... *Diospyros*

45.   Leaves (at least some) deeply cleft or lobed 46
      Leaves not lobed, but with sharply serrate to rounded teeth on margins 51

46.   Leaf stems typically longer than 2" 47
      Leaf stems usually 2" or less 49

47.   Leaves star-shaped; fruit a spherical "gum ball," with many sharp points .... *Liquidambar*
      Leaves not star-shaped; fruit otherwise 48

48.   Leaves with sharp (maple-like) lobes; fruit a fused ball of seeds (sycamore ball) .... *Platanus*
      Leaves with squared (truncated) terminal lobe; flowers tulip-like, orange-green; fruit a cluster of paddle-shaped seeds .... *Liriodendron*

49.   Leaves with rounded to sharp lobes; cluster of terminal buds; fruit an acorn .... *Quercus*
      Leaves variable in outline, oval to cordate to mitten-shaped, sometimes with 3 lobes 50

50.   Leaves with entire margins, fragrant when crushed; flowers in yellow-green clusters; fruit are blue drupes .... *Sassafras*
      Leaves with toothed margins; not fragrant; flowers greenish catkins; fruit an elongated multiple purple-black berry .... *Morus*

51.   Leaves with long (>2") stems (petioles) 52
      Leaf stems 2" or shorter 53

52.   Leaf stems round in cross-section; leaves heart-shaped, strongly toothed; flowers and fruit with attached leafy bract .... *Tilia*
      Leaf stems typically flat in cross-section; leaves round to triangular, toothed margins; buds often sticky; seeds with cottony floss .... *Populus*

53.   Leaf margins coarsely toothed (6 teeth or less/inch) 54
      Leaf margins finely toothed (>6 teeth/inch) 56

54.   Leaf margins wavy-toothed; cluster of terminal buds; fruit an acorn .... *Quercus*
      Leaf margins sharp-toothed; single end bud; fruit otherwise 55

55.   Leaf blades 5–10" long, very pointed tips, teeth very sharp; fruit an oval nut within a spiny bur .... *Castanea*
      Leaf blades 2–5" long, tips acute; end bud red-brown, slender, pointed, cigar-shaped; fruit a triangular nut in spiny husk .... *Fagus*

56.   Leaves almost round to heart-shaped 57
      Leaves oval to long-linear 58

57.   Leaves almost round (2–4"), margins wavy; yellow strap-like flowers in fall; fruit a woody capsule .... *Hamamelis*
      Leaves heart-shaped to variously lobed, with sharp tips and toothed margins; fruit elongated purple-black berries in

a cluster .... *Morus*

58. Leaves usually long-linear (occasionally elliptical), sharp tips and finely to remotely toothed margins; flowers in catkins; single bud scale .... *Salix*
    Leaves oval, finely and definitely toothed to notched; flowers and fruit various 59

59. Showy flowers (Rose Family), often perigynous (see glossary); fruit a drupe, fleshy berry or apple 60
    Flowers not showy; fruit various 63

60. Branches with sharp thorns or pointed twigs; leaves variously toothed or notched 61
    Branches without thorns; leaves uniformly serrate ... 62

61. Branches with pointed twigs; fruit a small apple ... *Pyrus*
    Branches with very sharp, sometimes branched, thorns; fruit small (<1/2"), fleshy, many seeded .... *Crataegus*

62. Leaves oval with obtuse, short pointed tips, finely toothed; fruit a several-seeded berry .... *Amelanchier*
    Leaves long oval with acute, sharper tips, minutely toothed; fruit a one-seeded drupe .... *Prunus*

63. Leaves strongly asymmetrical at base, slender with long tapered tips, weakly toothed; fruit a red-brown drupe .... *Celtis*
    Leaves not strongly asymmetrical, oval in outline, slightly to strongly toothed; fruit otherwise 64

64. Leaves generally thick with somewhat hairy surfaces; sharply and strongly toothed; flowers tiny, open before leafing; fruit papery, round-winged (samara).............. *Ulmus*
    Leaves thinner and smooth to slightly hairy; flowers larger, various; fruit not a samara 65

65. Leaves elliptical, shiny, short-pointed; flowers in umbrella-like clusters; fruit are red berries .... *Rhamnus* (*R. caroliniana*, only)
    Leaves not elliptical-shiny; flowers in catkins; fruit cylindrical, cone-like, bracted, or hop-like 66

66. Leaves somewhat round, teeth moderately coarse; twig

pith triangular; buds stalked .... *Alnus*
     Leaves more oval to triangular-pointed; fruit nutlets in varied structures 67

67. Leaves more coarsely toothed, oval to triangular-pointed; fruits multiple, cone-like; buds sessile .... *Betula*
     Leaves thin, finely doubly toothed; fruit bracted or hoplike 68

68. Leaves nearly glabrous; buds to 1/6"; bark gray, often ridged ("muscular"); fruit leafy-bracted nutlet .... *Carpinus*
     Leaves more pubescent; buds to 1/3"; bark brown, shreddy; fruit hop-like .... *Ostrya*

Division 2: Trees with Leaves Absent

1. Twigs with leaf scars opposite or whorled 2
     Twigs with leaf scars alternate 14

2. Branchlets with stubby compressed spur shoots (to 3/4" long), with crowded whorls of leaf scars near the tip .. 3
     Branchlets lacking spur shoots 4

3. Spur shoots short (to 3/8"), with crowded whorls of tiny leaf scars near the tips; buds small, somewhat pointed; reproductive structure a small (<1" long) scaly cone .... *Larix*
     Spur shoots longer (to 3/4"), stout, with fewer larger leaf scars in whorls near the tip; buds larger, squat, blunt; reproductive structure (female trees only) plum-like with odor .... *Ginkgo*

4. Twigs distinctly green or purplish green 5
     Twigs not green 6

5. Twigs green to purple green, with whitish bloom; buds medium, rounded, white hairy; fruit a double samara (maple "key") .... *Acer* (*A. negundo*, only)
     Twigs bright green, usually with longitudinal lines; buds small, pointed, smooth; fruit a scarlet capsule .... *Euonymus*

6. Bundle scars within leaf scar 1; buds small (to 1/8"), spherical, smooth; fruit a purple-black drupe .... *Forestiera*
     Bundle scars 3 or more; buds larger, not round; fruit various 7

7.   Bundle scars more than 7, usually arranged nearly in a circle or ellipse 8

Bundle scars 3 to 7, usually not arranged in a circle ... 9

8.   Pith chambered or hollow; leaf scars opposite; ellipse of bundle scars not quite closed; fruit a short capsule .... *Paulownia*

Pith solid; leaf scars usually in whorls of three; ellipse of bundle scars closed; fruit a long slender capsule ... *Catalpa*

9.   Leaf scars very broad (at least 1/3" across), conspicuous; end buds large, pointed, orange to brown; fruit large, nut-like ("buckeye") in a fleshy capsule .... *Aesculus*

Leaf scars smaller; end buds smaller, gray to brown; fruit a samara or drupe 10

10.   End bud broad, blunt, brown to nearly black; fruit a paddle-shaped samara .... *Fraxinus*

End bud pointed to rounded; fruit otherwise 11

11.   Buds with 2 exposed outer scales; twigs gray, gray-brown, or purplish; fruit a drupe 12

Buds with more than 2 exposed outer scales; twigs brown; fruit a double samara or a drupe 13

12.   Twigs purplish; leaf buds pointed, flower buds squat, rounded; fruits are small scarlet drupes ("berries") .... *Cornus* (*C. florida,* only)

Twigs gray or gray-brown; end buds chaffy, 3–5 times as long as broad; fruits are dark blue to black small drupes .... *Viburnum*

13.   Buds with several overlapping scales; fruits are double samaras (maple keys) .... *Acer*

Bud scales fewer, overlapping; ends of twigs taper to a spine; fruits black, berry-like, small (to 1/4"), bitter .... *Rhamnus* (*R. cathartica* only)

14.   Leaf scars very tiny and lacking bundle scars; "fruits" are woody, spherical cones; bark pale brown, fibrous .... *Taxodium*

Leaf scars larger, more conspicuous, bundle scars visible; fruit types many 15

15. Thorns or spines present 16
     Thorns typically absent (some species may have
     "thornless" individuals) 23

16. Spines in pairs, frequently at leaf scars 17
     Spines solitary, although sometimes branched 18

17. Spines straight, expanded and oval, at base; buds
     small, sunken in twig, without bud scales; fruit a dry
     pod(legume) .... *Robinia*
         Spines straight, much flattened at base; buds rounded,
     covered with red, woolly hairs; fruits are clusters of red-
     brown capsules .... *Zanthoxylum*

18. Spines scattered all along twigs and branches; pith
     large; fruit a black juicy berry .... *Aralia*
         Spines not usually scattered all along twigs and
     branches, but may occur on tree trunk; pith small; fruit
     otherwise 19

19. Thorns branched (to 6" long), very sharp and
     strong; fruit often a long (to 10") (or a shorter, more blunt)
     pod .... *Gleditsia*
         Thorns unbranched, less threatening; ends of
     branchlets often pointed, thorn-like 20

20. Branch tips very sharp-pointed 21
     Branch tips usually not sharp-pointed 22

21. Fruits are small apples; buds usually pointed at tip;
     twigs smooth, without scales .... *Pyrus*
         Fruits are small, scaly drupes; buds more
     rounded .... *Elaeagnus*

22. Terminal bud present, buds usually red; thorns slender,
     to 2" long; twigs somewhat straight; fruits small (to 3/4"
     diam.), apple-like, many seeded .... *Crataegus*
         Terminal bud absent; side buds tiny, red-brown;
     twigs zig-zag; fruits compound, very large (to 6"),
     spherical .... *Maclura*

23. Pith chambered, at least at twig nodes 24
     Pith solid, although sometimes with diaphragms 26

24. Leaf scars 3-lobed, usually with 3 groups of bundle scars; buds large (to 3/8") ....*Juglans*
  Leaf scars half-round, not 3-lobed; buds smaller 25

25. Pith chambered only at the nodes; stipular scars present; buds to 1/4" long, bud tips tightly appressed to twig; fruit a small drupe .... *Celtis*
  Pith chambered at nodes and between nodes; stipular scars absent; buds tiny (to 1/8"), rounded, not tightly pressed to twig; fruit an orange pulpy berry (to 1 1/2") .... *Diospyros*

26. Pith with diaphragms 27
  Pith solid without diaphragms 30

27. Buds without scales, brown-hairy, pointed at tip ....*Asimina*
  Buds with scales present 28

28. Bud scales 1 or 2; bundle scars 7 or more; stipule scars present 29
  Bud scales 3 or more; bundle scars 3; stipule scars absent; fruits are blue-black drupes .... *Nyssa*

29. Bud scale 1, hairy, large (to nearly 1" long); leaf scars crescent-shaped; fruit a large cluster of capsules ... *Magnolia*
  Bud scales 2, smooth, duck-bill shaped; leaf scars nearly round; fruit a cluster of samaras .... *Liriodendron*

30. Leaf scars forming a ring around the buds; bud with one scale; twigs smooth, light brown; bark of upper trunk mottled white and green .... *Platanus*
  Leaf scars not forming a ring round the buds; bark of trunk not mottled or white 31

31. Twigs aromatic or fragrant when cut or bruised 32
  Twigs not aromatic or fragrant 33

32. Twigs green, arching upward; buds greenish, smooth, pointed; fruits are deep blue drupes .... *Sassafras*
  Twigs orange to brown, usually straight, or drooping; buds brown, somewhat hairy, scales conspicuous; fruits are small, woody, cone-like (to 1 1/2" long) .... *Betula*

33. Buds stalked 34
    Buds not stalked (sessile) 35

34. Pith triangular in cross-section; buds with chaffy
scales; fruit small, woody, cone-like (to 3/4"); seeds tiny
nutlets .... *Alnus*
    Pith round in cross-section; buds naked, yellow-hairy;
fruits are small capsules (to 1/2"), brown, hairy; black seeds
small, shiny .... *Hamamelis*

35. Buds naked, pointed, lance-shaped (to 1/4"), very
hairy; fruits are round berries (to 1/3" diam.), red
and shiny .... *Rhamnus* (*R. caroliniana*, only)
    Buds with 1 or more scales; not pointed nor hairy;
fruit various 36

36. Buds with a single scale, small (to 1/8"), oblong,
smooth; twigs very slender, flexible to brittle; leaf scars
U-shaped; fruits are clusters of tiny capsules .... *Salix*
    Buds with 2 or more scales, buds larger; fruits
otherwise 37

37. Buds with 2 scales; usually asymmetrical, buds larger,
rounded to elongated; fruit a hard nut or nutlet with a leafy
bract 38
    Buds usually with 3 or more scales; fruits
otherwise 39

38. Buds with 2 scales, buds asymmetrical, rounded,
bright red in winter, green earlier; fruits are nutlets, round
(to 3/8"), several attached to leafy bract .... *Tilia*
    Buds with 2 scales, buds elongated red-brown to
sulfur yellow; fruit a hard nut in a husk
(hickory nut) .... *Carya* (in part)

39. Exposed bud scales 2–3; buds very tiny (<1/10");
twigs light red-brown; fruit a thin flat pod (legume) 3–6"
long .... *Albizia*
    Exposed bud scales 3 to several; buds larger 40

40. Pith star-shaped (stellate) in cross-section; fruit an
acorn; buds clustered at tips of twigs .... *Quercus*
    Pith round or nearly round (or sometimes lobed)

in cross-section; buds not clustered at twig tips; fruit otherwise 41

41. True terminal bud present 42
    True terminal bud not present 55

42. Terminal buds larger than rest of buds 43
    Terminal buds smaller than rest of buds or sometimes absent 53

43. Buds with 2–3 exposed scales, bundle traces usually 3 44
    Buds with 4 or more exposed scales; bundle traces usually 3 49

44. Twigs greenish to reddish; leaf scars elevated 45
    Twigs gray to brown; leaf scars usually not elevated ... 46

45. Twigs smooth, slightly upturned; leaf scars crescent-shaped; buds sharp-pointed with chestnut brown scales; fruits round, blue berries (to 1/3" diam.) on red stalks .... *Cornus* (*C. alternifolia*, only)
    Twigs dotted with pores, often zigzag; leaf scars triangular; buds spoon-shaped with dark red scales; fruits are spreading clusters of tiny 5-valved capsules .... *Oxydendrum*

46. Leaf scars half round to 3-lobed, bundle scars several to many; fruit are large nuts within a bur or husk 47
    Leaf scars more rounded, bundle scars few to several; fruit clustered white berries or a nutlet, in a paper husk ... 48

47. Leaf scars half round; twigs red-brown, slightly zigzag, usually smooth; fruit a large nut in a spiny bur .... *Castanea*
    Leaf scars usually 3-lobed; twigs gray to brown, straight, usually hairy; fruit a large nut in a thick husk .... *Carya* (in part)

48. Twigs rather stout with conspicuous pores (lenticels); buds rounded, hairy, purplish; fruits are clusters of white to creamy, smooth, shiny berries (to 3/8" diam.) **(This plant is extremely toxic—Do Not Touch!)** .... *Toxicodendron*
    Twigs more slender without conspicuous pores; buds not as above; fruits are nutlets within inflated paper

husks .... *Koelreuteria*

49. Buds oval, pointed, with conspicuous scales; leaf scars half-round; twigs red-brown, bitter to taste; fruit a drupe (cherry or plum) .... *Prunus*
    Buds slender to oval with conspicuous scales (twigs not bitter); fruit a nut, fused follicle, or cluster of capsules 50

50. Buds slender, at least 4 times as long as broad, smooth to the touch, scales often hairy on edges 51
    Buds more oval to cone-shaped, less than 4 times as long as broad, shiny, often sticky to the touch 52

51. Buds slender, narrow ("cigar-shaped"), smooth, light reddish-brown (to 3/4" long); bark of trees and twigs a smooth light gray; fruits are paired triangular nuts in a spiny husk .... *Fagus*
    Buds pointed, slender, dark red-brown (to 1/2" long); bark smooth, silvery early, then dark gray; fruits 1–3 seeded berries.. *Amelanchier*

52. Twigs slender, frequently with corky ridges; buds cone-shaped, sharp-pointed, shiny olive to red-brown scales; fruits are round clusters (to 1 1/2" diam.) of spiny follicles, pendant on a long stalk .... *Liquidambar*
    Twigs stout, brittle, without corky ridges; buds oval to lance-shaped, pointed, shiny brown to hairy on different species; fruits are clusters of small capsules with cottony hairs on seeds .... *Populus*

53. Twigs stout; pith medium, reddish-brown; leaf scars often 1/2" across; side buds superposed; fruit a thick, blunt pod (legume) .... *Gymnocladus*
    Twigs stout; pith large, tan to brown; leaf scars 1/4" across or larger; side buds not superposed; fruit otherwise 54

54. Leaf scars about 1/4" across; sap of twigs milky; twigs not malodorous; fruits are terminal clusters of dry, red berries... *Rhus*
    Leaf scars about 1/2" across; bundle scars 9; sap not milky; twigs malodorous when bruised; fruit are clusters of papery samaras .... *Ailanthus*

55. Leaf scars partially or completely encircling buds 56
    Leaf scars not encircling buds 57

56. Leaf scars completely encircling bud; fruit a thin pod (legume); bark of trees smooth, beech-like .... *Cladrastis*
    Leaf scars only partially encircling bud; fruit a round papery samara .... *Ptelea*

57. Leaf scars small, triangular, bordered by hair, bundle scars 3; buds very small, rounded; stipular scars absent; fruit a thin, narrow pod (legume) to 3 1/2" long .... *Cercis*
    Leaf scars small; stipular scars present; buds with 4 or more scales exposed 58

58. Sap milky; leaf scars half round; buds round to pointed, brown (to 1/4"); bundle scars not in groups of three .... *Morus*
    Sap not milky; bundle scars in groups of 3; buds scaly, pointed 59

59. Exposed bud scales about 12; buds about 1/8" long; bark smooth, dark, muscular; fruit nutlets with leafy bracts; male catkins usually not present in winter .... *Carpinus*
    Exposed bud scales about 6; buds usually >1/8" long; bark not smooth, muscular; fruit otherwise 60

60. Leaf scars and bundle scars slightly elevated; buds to 1/4"; bark brown and shreddy; fruit hop-like; male catkins present in winter .... *Ostrya*
    Leaf scars and bundle scars slightly depressed; bark usually furrowed; fruit a round samara .... *Ulmus*

Keys to Species of Diverse Genera

Division 1: Trees with Leaves Present

Key to the Genus *Acer*

The Maples—Summer Key

1. Leaves pinnately compound (3–5 leaflets); twigs green ....*Acer negundo*
    Leaves simple, palmately lobed; twigs brown or red ... 2

2.   Leafstalk with milky sap when broken ....*A. platanoides*
Leafstalk without milky sap 3

3.   Areas between leaf lobes typically V-shaped; lower leaf
surfaces silvery or white..4
Areas between leaf lobes typically U-shaped; lower leaf
surfaces green, or only paler than upper 5

4.   Leaves deeply divided and widely spreading;
branchlets arching upward, foul-smelling when bruised;
samaras 1 1/2" or more .... *A. saccharinum*
Leaves shallowly divided and narrower; branchlets
straight, not foul-smelling; samaras 1" or less ...... *A. rubrum*

5.   Leaves very dark green, edges drooping, typically
3-lobed ("goose-foot" shaped), lower surfaces hairy; small
leafy outgrowths (stipules) often present at base of
leafstalk .... *A. nigrum*
Leaves paler green, edges flat, typically 5-lobed,
lower surface usually sparsely or not hairy; stipules
absent .... *A. saccharum*

Key to the Genus *Aesculus*

The Buckeyes—Summer Key

1.   Leaflets mostly 7; buds very sticky; petals usually 5;
husk of nut very spiny .... *Aesculus hippocastanum*
Leaflets mostly 5; buds not sticky; petals usually 4;
husk spiny to smooth 2

2.   Husk somewhat spiny; stamens long, exerted
from flowers; twigs and leaves foul smelling when
crushed .... *A. glabra*
Husk smooth; stamens shorter, within flowers; twigs
and leaves without foul odor .... *A. flava*

Key to the Genus *Betula*[1]

The Birches—Summer Key

1.   Bark of trunk whitish, peeling in very thin strips; twigs
with prominent lenticels; leaves with no more than 7 pairs
of veins 2

Bark of trunk not white, peeling in strips; twigs with inconspicuous lenticels; leaves with 8 or more pairs of veins 3

2.   Leaves egg-shaped with pointed tips, rounded at base; branchlets not resinous glandular .... *B. papyrifera*
      Leaves delta-shaped with long tapering tips, squarish at base; branchlets densely resinous glandular .... *B. popufolia*

3.   Bark yellowish to silvery-gray or sometimes pale bronze, peeling in thin papery layers .... *B. allegheniensis*
      Bark reddish-tan to gray-brown, peeling in papery layers when young; dark brown and flaking in thick plates on old trees .... *B. nigra*

Key to the Genus *Carya*[2]

The Hickories—Summer Key

1.   Leaves with 11 or more leaflets, strongly aromatic when crushed; nut elongated, with thin husk
      ....*Carya illinoensis*
      Leaves mostly with fewer than 11 leaflets, not strongly aromatic; nut round to oval, husk thick or thin 2

2.   Leaflets typically 7 or 9 (occasionally 11); buds sulfur yellow with <6 adjoining scales; nut smooth, husk thin, splitting to middle .... *C. cordiformis*
      Leaflets typically 5 or 7 (occasionally 9, never 11); buds with 6 or more overlapping scales; nuts ridged, husk thin to thick 3

3.   Branchlets usually stout; terminal buds large (>1/2" long); nuts with husks >3/16" thick 4
      Branchlets usually slender; terminal buds small (1/4 to 1/2"); nuts with husks <3/16" thick 6

4.   Leaflets typically 5, teeth along margin with small tufts of hair; bark very shaggy; nuts strongly angled .... *C. ovata*
      Leaflets typically 7 (sometimes 9), teeth along margin without small tufts of hair 5

5.   Bark tight, not shaggy; trees typically on dry sites;

twigs and leaf surfaces densely hairy; nuts medium (to 1 1/4" long) .... *C. tomentosa*

Bark shaggy; especially on larger trees; trees typically on low ground; twigs and leaf surfaces smooth or sparsely hairy; nuts very large (1 1/4–2 1/2" long) .... *C. laciniosa*

6.   Leaflets mostly 5; bud scales without hair; bark tight, not peeling; nut mostly smooth, often with a short stem at base .... *C. glabra*

Leaflets mostly 7; bud scales with hairs at tip; bark peeling at maturity; nut angled above middle, without stem .... *C. ovalis*

Key to the Genus *Crataegus*[3]

The Hawthorns—Summer Key

1.   Leaves on flowering branches broadest below the middle, lobed, base broad and rounded, strongly pubescent; fruit pubescent .... *Crataegus mollis*

Leaves on flowering branches mostly obovate to oblong-elliptical, mostly lacking lobes; fruit usually smooth 2

2.   Leaves glossy on upper surfaces; fruit greenish to red, hard and dry, nutlets 1 to 2 .... *C. crus-galli*

Leaves dull on upper surfaces; fruit green to bright red, becoming succulent, nutlets 3 to 5 3

3.   Leaves thin, veins obscure above, sparingly hairy below; fruit 1/4 to 1/3" long, bright red to orange .... *C. viridis*

Leaves thick, veins impressed in upper surface, lower surface hairy throughout; fruit 1/3 to 3/4" long, green to red, with dots on surface .... *C. punctata*

Key to the Genus *Fraxinus*[4]

The Ashes—Summer Key

1.   Leaflets 7–11 (typically 9), sessile or very nearly so; fruits winged to the base 2

Leaflets 5–9 (typically 7), stalked, fruits not winged to the base 3

2.   Twigs usually strongly 4-angled; leaflets with very short stalks; flowers perfect .... *Fraxinus quadrangulata*
    Twigs round or nearly so; leaflets distinctly without stalks; at least some flowers unisexual .... *F. nigra*

3.   Leaves and branchlets smooth or nearly so; leaf scar usually deeply notched or V-shaped at top; fruit winged only to 1/3 of its length .... *F. americana*
    Leaves and branchlets hairy, at least when young; leaf scar shallowly notched or nearly straight across at top, fruit winged >1/3 of length 4

4.   Leaves very densely pubescent and white-appearing below; leaf scar sometimes notched at upper margin; fruit wing >1/3" wide .... *F. profunda*
    Leaves somewhat pubescent, pale green below; leaf scar nearly straight or convex at upper margin; fruit wing <1/3" wide .... *F. pennsylvanica*

Key to the Genus *Pinus*[5]

The Pines—Summer and Winter Key

1.   Leaves in clusters of 5, thin, soft, flexible (3–5" long); woody cones slender (finger-shaped), without prickles, 3–7" long .... *Pinus strobus*
    Leaves in clusters of 2–3, stiffer (1 1/2–6" long); cones woody, various-shaped, often armed with prickles 2

2.   Leaves in clusters of 2 or 3, flexible, usually less than 5" long; cone scale with a small spine .... *P. echinata*
    Leaves in clusters of 2, flexible or stiff, widely variable in length; cones armed, or not 3

3.   Most or all leaves >3", rarely twisted 4
    Most or all leaves <3" long, sometimes twisted 5

4.   Leaves flexible, slender, snap easily when bent double; cone scales without prickles .... *P. resinosa*
    Leaves very stiff, rather stout, do not snap when bent double; cone scales with prickles .... *P. nigra*

5.   Cone scales with a terminal spine; twigs with a waxy bloom, purplish; leaves 1 1/2–2 3/4" .... *P. virginiana*

Cone scales without terminal spines; twigs without a waxy bloom, yellow-brown to orange; leaves often twisted 3/4–3" 6

6.   Leaves <1 1/2" long; cones often erect; upper branches gray to brown .... *P. banksiana*
       Leaves 1 1/4–2 3/4" long; cones pendant; upper branches orange to red-orange .... *P. sylvestris*

Key to the Genus *Populus*[6]

The Cottonwoods and Aspens—Summer Key

1.   Leaves with lower surfaces covered with white, felt-like, dense hair, edges with 3–5 large teeth or small lobes .... *Populus alba*
       Leaves green, not white-hairy below when mature, edges with many teeth 2

2.   Leaf stalks round, often grooved, easily rolled between fingers; leaf bases slightly heart-shaped .... *P. heterophylla*
       Leafstalks flattened, difficult to roll between fingers; leaf bases rounded to flat 3

3.   Leaves basically triangular in shape, long-pointed, coarsely toothed; buds typically sticky 4
       Leaves basically oval or spherical, coarse or fine-toothed; buds typically not sticky 5

4.   Leaves coarsely toothed, margins usually densely hairy; two glands near top of leafstalk; native trees with spreading crowns .... *P. deltoides*
       Leaves less coarsely toothed, margins lacking hair; glands lacking; introduced trees with tall, columnar crowns .... *P. nigra* var. *italica*

5.   Leaves with up to 15 coarse teeth along edges, mostly 2 to 5" long.. *P. grandidentata*
       Leaves with 20 or more fine teeth along edges; mostly 1 1/4 to 4" long .... *P. tremuloides*

Key to the Genus *Prunus*

The Cherries and Plums—Summer Key

1.   Margins of leaves with sharp teeth, teeth without glands; fruit a round plum, red, to nearly 1" diameter .... *Prunus americana*
     Margins of leaves with blunt or rounded teeth; fruit various 2

2.   Leaves usually less than twice as long as wide; branchlets often with thorns; fruit an oval plum, to 3/4" long .... *P. nigra*
     Leaves mostly greater than twice as long as wide; branchlets usually without thorns 3

3.   Leafstalks hairy; leaves irregularly toothed; fruit plums to 1" diam., red to yellow .... *P. hortulana*
     Leafstalks mostly not hairy; leaves regularly toothed; fruit cherries to 3/8" diam. 4

4.   Leaves with dense red-brown hairs on midrib below; paired small glands at upper end of leafstalk; fruit in clusters of 10 or more, black .... *P. serotina*
     Leaves without hair on midrib, leaves wrinkled; glands absent from leafstalk; fruit in clusters of 2 to 7, red   .... *P. pensylvanica*

Key to the Genus *Quercus*

The Oaks—Summer Key

1.   Leaf margins without teeth, about 1" wide, hairy below; acorns small (<1 1/2"), oval, cup covers <1/2 of nut   .... *Quercus imbricaria*
     Leaf margins variously and prominently toothed or lobed 2

2.   Lobes or teeth of the leaves rounded to acute, but never bristle-tipped (white oak group); acorns mature in 1 season 3
     Lobes of the leaves bristle-tipped (red or black oak group); acorns mature in 2 seasons 10

3.   Leaves distinctly and often deeply lobed, the lobes 2–5 on each side 4

   Leaves coarsely toothed, with 3–14 teeth on each side, the leaf notches extending <1/3 of the way to the mid-vein 7

4.   Lower leaf surface devoid of hair at maturity; acorn to 1", cup deeply saucer-shaped, covers 1/4–1/3 of nut .... *Q. alba*

   Lower leaf surface hairy to the touch at maturity 5

5.   Upper three lobes squarish, forming a cross; twigs hairy; acorn small <2/3" long, cup bowl-shaped, covers 1/2 of nut .... *Q. stellata*

   Leaves without three squarish lobes at upper end; twigs smooth, or nearly so; acorn larger, cup deeper 6

6.   Upper lobe of large leaf very broad, notch often cut nearly to midrib; twigs often with corky ridges; acorns very large (to 2"), cup encloses 1/2–2/3 of nut, conspicuously fringed at top .... *Q. macrocarpa*

   Leaf lobes of narrower leaf more regularly spaced; corky ridges absent; acorn smaller (to 1"), globe-shaped, cup deep, covers 2/3 to nearly all of nut .... *Q. lyrata*

7.   Leaves with silvery-white lower surface due to velvety hairs; acorns on long, slender stalks (1–2 3/4"), the acorn stalks longer than leafstalks .... *Q. bicolor*

   Leaves with lower surfaces green or pale, but not silvery white; acorns sessile or on stalks shorter than leafstalks 8

8.   Leaves usually with sharp-pointed teeth, tiny gland at each tooth tip; acorns small (<1" long), nearly black at maturity.. *Q. muehlenbergii*

   Leaves usually with somewhat rounded teeth, glands not present; acorns larger (to 1 1/2" long), rich brown at maturity... 9

9.   Leaves velvety-hairy below; acorn cup 1" or more across; trees of low ground; bark like white oak .... *Q. michauxii*

   Leaves smooth below or sparsely hairy; acorn cup

<1" across; trees of high ground; bark more like red oak .... *Q. prinus*

10. Leaves distinctly broader above the middle, leaf base rounded; twigs and buds densely hairy; acorn to 2/3" long, cup covers 1/3–1/2" of nut; occurs on dry ridges .... *Q. marilandica*
    Leaves broadest at or below middle; twigs and buds usually not densely hairy; acorns various 11

11. Leaves hairy throughout lower leaf surface 12
    Leaves smooth on the lower leaf surface or hairy only next to the veins; terminal buds oval, scales usually not hairy 14

12. Lower leaf surface with rusty hairs; base of leaf blade not broadly rounded; buds 4-angled, densely hairy; acorns <3/4" long, cup covers 1/2 of nut, cup scales loose, fringe nut .... *Q. velutina*
    Lower leaf surface with gray hairs; base of leaf blade broadly rounded (bell-shaped); buds less angled and hairy; acorn cup not fringe-scaled 13

13. Leaves with 3–5 more or less unequal lobes, the uppermost lobes often strongly curved (scythe-shaped); usually on dry sites .... *Q. falcata*
    Leaves with 5–11 more or less equal lobes, the uppermost lobes not strongly curved; usually in bottom-lands .... *Q. pagoda*

14. Leaves shallowly lobed, divided less than halfway to middle; lower leaf surface hairy only in vein axils; acorn large (to 1 1/4") with shallow, saucer-shaped cup covering only 1/5–1/4 of nut .... *Q. rubra*
    Leaves deeply divided more than halfway to middle; acorns smaller (<1") 15

15. Acorn cup shallowly saucer-shaped, covering about 1/4 of nut 16
    Acorn cup top-shaped, covering about 1/2 of nut 17

16. Acorn small, usually striped, cup to 1/2" across; leaves with 2–4 lobes per side, nearly at right angles to leafstalk, leaf notches spreading .... *Q. palustris*

Acorn larger, not striped, cup >1/2" across; leaves with 3–5 lobes per side, angled to leafstalk, leaf notches closing .... *Q. shumardii*

17. Cup scales hairy and dull; kernel of nut yellow or orange and very bitter; inner bark yellowish or orange .... *Q. ellipsoidalis*[7]
Cup scales nearly glabrous and shiny; kernel white and not very bitter; inner bark gray or reddish .... *Q. coccinea*[8]

Key to the Genus *Rhus*

The Sumacs—Summer Key

1. Leaflets entire; leaf stalk winged .... *Rhus copallinum*
   Leaflets toothed; leafstalk not winged 2

2. Leaves, leafstalks, and twigs smooth .... *R. glabra*
   Twigs and leafstalks densely hairy .... *R. typhina*

Key to the Genus *Salix*

The Willows—Summer Key

1. Leaves 1–4 times as long as wide, typically widest at or above the middle 2
   Leaves 5–15 times as long as wide, typically broadest near the base, lanceolate 3

2. Leaf edges sparsely toothed, leaves oval to elliptical; buds large (1/4–3/8"), red-brown; flowers silky gray, open before leafing .... *Salix discolor*
   Leaf edges with many fine teeth, 3–4 times as long as wide (peach-leaved); buds small (<1/4"), brown, shiny; flowers with leaves, smaller .... *S. amygdaloides*

3. Leaves rounded at base, about 8 times as long as wide, green on both sides; twigs dark green, spreading ...... *S. nigra*
   Leaves taper gradually to a narrow base, about 5–15 times as long as wide 4

4. Leaves 10–15 times as long as wide, bright green above, paler below; twigs yellowish to orange, drooping .... *S. babylonica*

Leaves 5–8 times as long as wide, green above, pale or whitish below; twigs not yellowish, spreading 5

5.   Leaves dark green with white, silky hairs below when young, teeth on edges 20–25 per inch; twigs tough, do not break easily, spreading from branches at narrow angles (30–45°) .... *S. alba*
        Leaves gray-green, not white-hairy below, teeth 15–20 per inch; twigs fragile, break easily, spreading from branches at wide angles (60–90°) .... *S. fragilis*

Key to the Genus *Ulmus*[9]

The Elms—Summer Key

1.   Twigs (at least some) with corky wings or ridges 2
        Twigs without corky wings or ridges 3

2.   Leaves 3–5 1/2" long, leafstalks >1/8" long; buds ovate, blunt; corky wings thinner .... *Ulmus thomasii*
        Leaves 1 1/2–2 3/4" long, leafstalks <1/8"; buds small, narrow, very sharp pointed; corky wings expanded, conspicuous .... *U. alata*

3.   Leaves very rough above; flowers not stalked; fruit not hairy; inner bark gluey .... *U. rubra*
        Leaves smooth or softer to the touch; flowers stalked; fruit usually hairy; inner bark not gluey 4

4.   Leaf edges doubly toothed, 3–4 1/2" long .... *U. americana*
        Leaf edges singly toothed, 1 1/2–2 1/2" long .... *U. pumila*

Key to the Genus *Viburnum*

The Black Haws—Summer Key

1.   Leafstalks winged; leaves ovate, long-pointed tips; typical of boggy or wet places .... *Viburnum lentago*
        Leafstalks not winged or narrowly expanded; leaves oval to obovate, short pointed or rounded tips; typical of dryer habitats 2

2.   Leaves oval; merely acute or obtuse at tip; absence of rusty hairs on leaves and buds .... *V. prunifolium*
       Leaves oval to obovate, rounded at tip; buds and leaves with rusty hairs .... *V. rufidulum*

NOTES

1. *Betula populifolia* formerly occurred naturally in Indiana, and may still occur; planted individuals can be found within the state. See "Species Excluded" on page 337.

2. *Carya pallida* and *C. texana* are also found occasionally in southwestern Indiana. See "Species Descriptions" on page 75 and "Species Excluded," page 337.

3. The genus *Crataegus* is notorious for being a difficult genus to identify its component taxa accurately and consistently. Hybridization and other forms of genetic mixing are widespread and continuing among many populations, resulting in confusing characteristics, and often defying the efforts of even professional taxonomists to identify individuals with confidence. The phenotypic variation that is evident in wild populations becomes even more of a problem when cultivated and ornamental forms (of which there are many) are considered.

A brief summary of previous efforts to classify the group follows: Sargent (1949 edition) listed 199 species and varieties for North America; Gleason and Cronquist (1991 edition) includes 177 species, varieties and hybrids for NE U.S. and adjacent Canada; Cope (2001) lists 28 species for Eastern U.S.; Swink and Wilhelm (1994) listed 12 species for the Chicago Region; Mohlenbrock (1973) included 7 species for Illinois. For Indiana, Deam's 1931 edition (4th) of *Trees of Indiana* included 23 species and 1 variety; whereas his 1953 edition (5th) listed only 4 species and 1 variety.

In this manual, we have elected to feature only the 4 species recognized by Deam (1953).

4. *Fraxinus biltmoreana, F. lanceolata,* and *F. pennsylvanica* var. *subintegerima* of some manuals have been found in Indiana, and may still occur here according to that nomenclature. See "Species Excluded" on page 337.

5. A number of other species of pines (both from elsewhere in the U.S. and exotic species) are planted in Indiana, both ornamentally and for forestry purposes. We chose to feature only those species that are likely to be encountered.

6. *Populus balsamifera* previously occurred in Indiana and may possibly still be present. See "Species Excluded" on page 337.

7. Some authors feel that *Quercus ellipsoidalis* represents a northern expression of *Q. coccinea* (e.g., Voss, Swink, and Wilhelm, and Wm. Overlease), and that a continuous intergradation occurs among populations of these taxa within Indiana. We elected to retain the traditional separation of species (as suggested by Jensen), but recognize that combining the two populations may be the most accurate interpretation of their actual taxonomic position.

8. See note 7.

9. *Ulmus parvifolia,* which is closely similar to *U. pumila* (both are exotic species), has undoubtedly been planted at many locations in Indiana, and may be encountered occasionally. We chose to recognize *U. pumila* as the common and widespread small-leafed exotic species of elm in the state.

Division 2: Trees with Leaves Absent

Key to the Genus *Acer*

The Maples—Winter Key

1.   Twigs green; buds covered with dense white hairs...*Acer negundo*
      Twigs gray, brown, or red; buds red, brown, or black .. 2

2.   End buds >1/4" long; sap milky...............*A. platanoides*
      End buds <1/4" long; sap clear 3

3.   Buds brown to black; bud scales 6 or more 4
      Buds reddish to orange-red; bud scales about 4 5

4.   Twigs shiny, buff to red-brown; buds nearly black, nearly smooth ....*A. saccharum*
      Twigs dull, straw-colored; pores prominent; buds pale brown, hairy ....*A. nigrum*

5.   Twigs bright chestnut brown; have a disagreeable odor when bruised ....*A. saccharinum*
      Twigs red and lustrous; no rank odor when bruised..*A. rubrum*

Key to the Genus *Aesculus*

The Buckeyes—Winter Key

1.　Buds large, nearly black; sticky to the touch .... *Aesculus hippocastanum*
　　Buds medium, brown to orange; smooth to the touch 2

2.　Bark of twig foul-smelling when bruised; fruit husk spiny; bark of trunk rough, soft, and corky .... *A. glabra*
　　Bark of twig not foul-smelling; fruit husk smooth; bark of trunk smooth and firm .... *A. flava*

Key to the Genus *Betula*[1]

The Birches—Winter Key

1.　Twigs greenish-brown, aromatic (have a mild wintergreen fragrance); bark on mature trees yellowish to bronze, scaly; buds hairy, appressed to twigs ...... *Betula alleghaniensis*
　　Twigs gray or reddish-brown, not aromatic; bark not yellowish; buds not appressed 2

2.　Bark on mature trees grayish-white, usually not peeling; twigs gray, with warty pores; buds tapering from center toward both ends .... *B. populifolia*
　　Bark on mature trees peeling in thin strips; twigs brown to black; buds tapering from base to tip 3

3.　Twigs reddish-brown, somewhat hairy; bark on mature trees salmon-pink, darker on very old trees .... *B. nigra*
　　Twigs nearly black, usually smooth; bark on older trees chalky white .... *B. papyrifera*

Key to the Genus *Carya*[2]

The Hickories—Winter Key

1.　Bud scales in pairs, not overlapping 2
　　Bud scales more than two, overlapping 3

2.　Buds sulfur-yellow with bran-like scales; nut rounded, husk thin, kernel very bitter .... *Carya cordiformis*

Buds covered with brownish hairs; nut elongated, husk thin, kernel sweet, tasty .... *C. illinoensis*

3.   Twigs stout; end bud large (>3/8" long), usually hairy; fruit with a thick husk 4

    Twigs slender; end bud smaller (<3/8" long), usually smooth; fruit with thin husk 6

4.   Outer bud scales falling early, bud smooth and neat; twigs densely hairy; bark not shaggy .... *C. tomentosa*

    Outer bud scales persistent, bud shaggy in appearance, twigs somewhat hairy; bark very shaggy 5

5.   Twigs dark reddish-brown; fruit 1–1 1/4" long .... *C. ovata*

    Twigs orange-brown or light brown; fruit 1 1/4–2 1/4" long .... *C. laciniosa*

6.   End bud essentially hairless, except for inner bud scales 7

    End bud covered with silvery scales .... *C. pallida*

7.   Bark smooth or furrowed, not scaly; fruit husk splitting only part way .... *C. glabra*

    Bark scaly, especially on larger trees; fruit husk splitting to base .... *C. ovalis*

Key to the Genus *Crataegus*[3]

The Hawthorns—Winter Key

1.   Trees of low alluvial ground, occurs only in far southwestern Indiana; bark thin, pale gray, peeling over orange-brown inner bark; often thornless or sparingly thorny.. *Crataegus viridis*

    Trees usually on drier sites, more widespread throughout Indiana; bark gray or gray-brown, thicker, becoming rough or scaly with age; typically rather thorny 2

2.   Bark on old trees thick and deeply ridged and fissured; branches stout, thorny. Reaches larger size than succeeding species .... *C. mollis*

    Bark on older trees not deeply ridged or fissured, but often scaly. Smaller trees 3

3.   Branchlets rather stout, often hairy when young; bark on older trees gray-brown, slightly scaly; usually armed with long, slender, nearly straight thorns, sometimes thornless .... *C. punctata*

Branchlets slender, smooth; bark on older boles scaly, smooth and dark gray on branches; usually densely branched and thorny, thorns long, curved .... *C. crus–galli*

Key to the Genus *Fraxinus*

The Ashes—Winter Key

1.   Twigs rather sharply 4-angled ... *Fraxinus quadrangulata*
Twigs rounded 2

2.   Buds black or very nearly so .... *F. nigra*
Buds brown or red brown 3

3.   Leaf scars nearly straight across the top .... *F. pennsylvanica*
Leaf scars notched at the top 4

4.   Twigs with conspicuous large lenticels; trunks usually swollen at the base; typical of very wet sites .... *F. profunda*
Twigs with inconspicuous lenticels; trunks not swollen at base; typical of moist sites .... *F. americana*

Key to the Genus *Populus*

The Poplars—Winter Key

1.   Twigs densely hairy; bark smooth, light gray or white to silvery .... *Populus alba*
Twigs smooth or sparsely hairy; bark gray, grayish-green, to reddish-brown 2

2.   Side buds with more than 4 exposed scales 3
Side buds with 3–4 exposed scales 4

3.   Twigs and buds smooth, shiny brown, appear varnished; bark light green to nearly white and birch-like .... *P. tremuloides*
Twigs and buds sparsely silky, dull gray to yellow-

brown; bark smooth, grayish-green .... *P. grandidentata*

4. End buds mostly 1/2" or longer 5
    End buds mostly <1/2" 6

5. Twigs reddish-brown; buds quite sticky, very fragrant
    when crushed .... *P. balsamifera*[4]
    Twigs yellowish-brown; buds widest near the middle,
    slightly sticky, not fragrant .... *P. deltoides*

6. Buds short and broad; crown spreading; native tree of
    swamps and wet sites .... *P. heterophylla*
    Buds slender and tapering; tall, columnar crowns;
    introduced tree usually planted .... *P. nigra* var. *italica*

Key to the Genus *Prunus*

The Plums and Cherries—Winter Key

1. Terminal (end) bud present 2
    Terminal bud absent or soon deciduous 4

2. Buds small (to 3/16"), scales red-brown, clustered at
    the tips of all shoots; twigs usually less than 1/16" thick;
    pith brown .... *Prunus pennsylvanica*
    Buds larger (to 1/2"), scales brown, not clustered at
    tips; twigs >1/16" thick, have disagreeable taste and odor
    when bruised; pith white 3

3. Buds usually 1/4" or less, scales chestnut-brown,
    keeled; frequently large trees; bark nearly black and very
    scaly when mature .... *P. serotina*
    Buds 1/4–1/2", scales dull brown, grayish on
    edges; large shrubs or small trees; bark nearly
    smooth .... *P. virginiana*

4. Branchlets unarmed with spine-tipped twigs; buds
    large (1/4–1/2" long); medium-sized trees .... *P. hortulana*
    Branchlets armed with spine-tipped twigs; buds
    smaller (<1/4" long); large shrubs to small trees 5

5. Bark of older trees separating into plates; surface
    of fruit stone usually smooth; widespread throughout
    Indiana .... *P. americana*[5]

Bark of older trees scaly; surface of fruit pit roughened; local in northern and central Indiana .... *P. nigra*[5]

Key to the Genus *Quercus*

The Oaks—Winter Key

1. Largest terminal bud at least 1/4" long 2
    Largest terminal bud <1/4" long 10

2. Buds distinctly angled in cross section 3
    Buds circular or barely angled in cross section 8

3. Buds smooth, dull straw color .... *Quercus shumardii*
    Buds hairy, at least at tip, gray to dark red 4

4. Buds hairy all over 5
    Buds hairy only at the tip 7

5. Buds with gray hairs; twigs often shiny ...... *Q. velutina*
    Buds with rusty or brown hairs; twigs usually dull 6

6. Buds with rusty hairs; twigs often pubescent; acorn small, about 1/2" across .... *Q. marilandica*
    Buds with brown hairs; twigs smooth; acorn large, to 1 1/4" long .... *Q. michauxii*

7. Buds dark red-brown; twigs smooth, acorn about 3/4" across .... *Q. coccinea*
    Buds light red-brown; twigs often somewhat hairy; acorn about 1/2" across .... *Q. pagoda*

8. Buds and twigs orange-brown; buds slender, acute .... *Q. prinus*
    Buds and twigs red, red-brown, or gray-brown 9

9. Buds light red to light red-brown; acorn very large and robust (to 1 1/2" long), cup shallow (covers <1/4 of nut) .... *Q. rubra*
    Buds dark red to dark-red brown; acorn small (to 1/2" long), cup deeper (covers 1/3 of nut) .... *Q. falcata*

10. Buds obtuse, more or less rounded 11
    Buds mostly pointed at the tip 16

11. Twigs shiny, red or reddish-brown 12
     Twigs dull, gray to yellow-brown to purplish 13

12. Buds rounded, not angular; acorns medium (to 3/4"
long), cup covers to 1/4 of nut .... *Q. alba*
     Buds somewhat angular; acorns small (to 1/2" across),
cup bowl-shaped, covers 1/3–1/2 of nut .... *Q. ellipsoidalis*

13. Twigs and buds hairy 14
     Twigs and buds smooth or nearly so 15

14. Buds red-brown; acorn 1/2–2/3" across, cup covers 1/2
of nut .... *Q. stellata*
     Buds gray to gray-brown; acorn huge
(to 1 3/4" across), cup covers 1/2 of nut, cup fringed
at top .... *Q. macrocarpa*

15. Twigs purplish with a whitish coating; acorn large (to
1 1/4" long) on a long (to 1 1/2") stalk, cup covers 1/3
of nut .... *Q. bicolor*
     Twigs gray to yellow-brown; acorn round (to 1" diam.),
cup encloses almost all of nut .... *Q. lyrata*

16. Buds and twigs brown to orange-brown; twigs usually
smooth; acorn to 3/4" long, chestnut-colored to nearly
black, cup covers to 1/2 of nut .... *Q. muehlenbergii*
     Buds and twigs red to red-brown to gray-brown; twigs
smooth or hairy 17

17. Scales of buds hairy; twigs light to dark brown; buds
gray-brown; acorns small (<1/2" long), cup covers 1/3–1/2
of nut; persistent leaves not lobed .... *Q. imbricaria*
     Scales of buds smooth or nearly so; twigs brown to
reddish; acorns small (<1/2" long), cup saucer-shaped,
encloses only the base; persistent leaves lobed nearly to
mid-rib .... *Q. palustris*

Key to the Genus *Rhus*

The Sumacs—Winter Key

1.   Leaf-scars U-shaped; twigs medium-stout, round,
sparingly hairy .... *Rhus copallina*

Leaf-scars horseshoe-shaped, nearly encircling the buds; twigs stout, pith large 2

2.   Twigs velvety-hairy, rounded .... *R. typhina*
      Twigs smooth, somewhat 3-sided .... *R. glabra*

Key to the Genus *Salix*

The Willows—Winter Key

1.   Primarily shrubs to small trees; twigs stoutish, red-brown to dark brown, mostly smooth; buds large for willows (to 3/8" long), single scale red-brown .. *Salix discolor*
      Primarily small to medium trees; twigs mostly slender, yellowish to nearly black; buds usually small (except *S. fragilis*), single scale yellow to brown 2

2.   Twigs long, pendulous (weeping), yellow to orangeish in winter .... *S. babylonica*
      Twigs not markedly pendulous 3

3.   Twigs yellow or yellowish, smooth; buds small (<1/6"), yellow .... *S. amygdaloides*
      Twigs yellowish-greenish-brownish; range of bud sizes 4

4.   Native trees, largest of the willows; twigs greenish-yellow; bark dark brown to nearly black; buds to 1/4" long; reddish brown .... *S. nigra*
      Introduced trees, smaller; twigs greenish-yellow to red-brown; range of bud size 5

5.   Twigs very brittle at base; bark gray, rough; buds mid-size (to 1/3") .... *S. fragilis*
      Twigs not brittle; bark gray-brown, ridged; buds small (to 1/8") .... *S. alba*

Key to the Genus *Ulmus*

The Elms—Winter Key

1.   Buds less than 1/8" long; branches often very twiggy near the ends .... *Ulmus pumila*

Buds more than 1/8"; branches usually less twiggy 2

2. Twigs (at least some) with corky outgrowths 3
Twigs without corky wings 4

3. Buds about 1/8" long; corky wings pronounced on
most branchlets .... *U. alata*
Buds about 1/4" long; irregular corky growths on some
branchlets .... *U. thomasii*

4. Buds about 1/4" long, rusty-hairy, very dark brown to
nearly black; twigs rough, ashy gray in color .... *U. rubra*
Buds 1/8–3/16" long, smooth to sparingly hairy,
brown, twigs usually smooth, brownish .... *U. americana*

Key to the Genus *Viburnum*

The Black Haws—Winter Key

1. Buds long and narrow (to 3/4"), at least 5 times longer
than broad .... *Viburnum lentago*
Buds shorter (to 1/2"), at most 3 times longer than
broad 2

2. Buds smooth to scurfy (covered with bran-like scales)
V. prunifolium
Buds rusty, hairy .... *V. rufidulum*

NOTES

1. *Betula populifolia* is considered to be extirpated from natural
communities in Indiana, but may be encountered as a planted tree.

2. *Carya texana* may be encountered on occasion in southwestern Indiana on dry sites. See page 339.

3. See page 64 note 3 for further explanation about the genus
Crataegus.

4. *P. balsamifera* is extirpated from Indiana, or very nearly so. It
is included within this key for completeness.

5. These species are difficult to separate consistently in winter
condition.

Section 1:
101 Native Species of Tree Size

Pinaceae
Tsuga canadensis, Eastern hemlock
Larix laricina, Tamarack
Pinus strobus, Eastern white pine
P. banksiana, Jack-pine
P. virginiana, Virginia-pine, Scrub-pine

Taxodiaceae
Taxodium distichum, Bald cypress

Cupressaceae
Thuja occidentalis, Northern white cedar
Juniperus virginiana, Eastern red cedar

Magnoliaceae
Magnolia acuminata, Cucumber-tree
M. tripetala, Umbrella-tree
Liriodendron tulipifera, Tulip-tree

Annonaceae
Asimina triloba, Pawpaw

Lauraceae
Sassafras albidum, Sassafras

Platanaceae
Platanus occidentalis, American sycamore

Hamamelidaceae
Liquidambar styraciflua, Sweet gum, Red gum

Ulmaceae
Ulmus americana, American elm
U. rubra, Slippery elm
U. thomasii, Rock-elm, Cork elm
U. alata, Winged elm
Celtis laevigata, Southern hackberry, Sugarberry
C. occidentalis, Northern hackberry, Hackberry

Moraceae

Maclura pomifera, Osage-orange
Morus rubra, Red mulberry

Juglandaceae

Juglans cinerea, Butternut
J. nigra, Black walnut
Carya illinoensis, Pecan
C. cordiformis, Bitternut-hickory
C. laciniosa, Shellbark-hickory
C. tomentosa, Mockernut-hickory
C. ovata, Shagbark-hickory
C. ovalis, Red-hickory
C. glabra, Pignut-hickory
C. pallida, Pale hickory

Fagaceae

Fagus grandifolia, American beech
Castanea dentata, American chestnut
Quercus alba, White oak
Q. stellata, Post-oak
Q. lyrata, Overcup-oak
Q. macrocarpa, Bur-oak
Q. bicolor, Swamp white oak
Q. michauxii, Swamp chestnut-oak
Q. prinus, Rock chestnut-oak
Q. muehlenbergii, Chinkapin-oak
Q. imbricaria, Shingle-oak
Q. marilandica, Black-jack oak
Q. pagoda, Cherrybark-oak
Q. falcata, Southern red oak
Q. velutina, Black oak
Q. rubra, Northern red oak
Q. palustris, Pin-oak
Q. shumardii, Shumard oak
Q. ellipsoidalis, Northern pin-oak
Q. coccinea, Scarlet oak

Betulaceae

Ostrya virginiana, Hop-hornbeam
Carpinus caroliniana, Hornbeam, Blue beech
Betula alleghaniensis, Yellow birch
B. nigra, River birch
B. papyrifera, White birch, Paper birch

Tiliaceae

Tilia americana, Basswood

Salicaceae

Populus grandidentata, Big-toothed aspen, Large-toothed aspen

P. tremuloides, Quaking aspen

P. heterophylla, Swamp-cottonwood

P. deltoides, Cottonwood

Salix nigra, Black willow

Ericaceae

Oxydendrum arboreum, Sourwood

Ebenaceae

Diospyros virginiana, Persimmon

Rosaceae

Prunus serotina, Wild black cherry

P. pennsylvanica, Pin-cherry, Fire cherry

P. americana, American wild plum

P. hortulana, Hortulan plum

P. nigra, Canada-plum

Pyrus coronaria, Sweet crabapple, American crabapple

P. ioensis, Prairie crabapple

Crataegus crus-galli, Cockspur-thorn

C. viridis, Green hawthorn

C. punctata, Dotted hawthorn

C. mollis, Downy hawthorn

Amelanchier arborea, Downy serviceberry

A. laevis, Smooth serviceberry, Allegheny shadbush

Caesalpiniaceae

Cercis canadensis, Redbud

Gleditsia triacanthos, Honey-locust

G. aquatica, Water-locust

Gymnocladus dioica, Kentucky coffee-tree

Fabaceae

Cladrastis lutea, Yellow-wood

Robinia pseudoacacia, Black locust

Cornaceae

Cornus florida, Flowering dogwood

Nyssa sylvatica, Black gum

Hippocastanaceae
Aesculus glabra, Ohio-buckeye
A. flava, Yellow buckeye

Aceraceae
Acer saccharum, Sugar-maple
A. nigrum, Black maple
A. rubrum, Red maple
A. saccharinum, Silver-maple
A. negundo, Boxelder

Anacardiaceae
Rhus typhina, Staghorn-sumac

Oleaceae
Fraxinus americana, White ash
F. pennsylvanica, Green ash
F. profunda, Pumpkin-ash
F. nigra, Black ash
F. quadrangulata, Blue ash

Bignoniaceae
Catalpa speciosa, Northern catalpa

EASTERN HEMLOCK

Tsuga canadensis (L.) Carrière.

Distinguishing Features: **Medium to large evergreen tree** (to 70 feet, 2+ feet diam.) with a broad, open, pyramid-shaped crown and a drooping leader. **Needle leaves borne singly (<3/4 inch long), have two white lines below. Cones at branch tips are about 3/4 inch long.**

Leaves: Needle-like, flattened, very numerous, borne singly on short stalks; appear to be in two rows on either side of twig; blunt or notched at tip. Dark, glossy yellow-green, and grooved above; two narrow lengthwise white bands below. Needles soft to touch.

Bark: Scaly on young trees, becoming deeply furrowed and ridged with maturity; gray to red-brown.

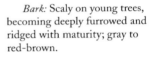

Twigs/Buds: **Twigs** slender, **spreading to drooping** (especially the topmost branch); pale brown and fuzzy early, then gray and smooth. Buds oval, red-brown, very small (1/16 inch).

Flowers/Fruit/Seeds: Cones unisexual on same tree. Male cones very small (to 3/8 inch), yellowish, near branch tips in spring. Female cones leathery, green, larger, also near branch tips. **"Fruit," a rich-brown woody cone (to 3/4 inch long),** matures in autumn; winged seeds (1/8 inch) are wind-scattered in winter.

Habitat: On steep, rocky bluffs (usually sandstone), and in deep wooded ravines where cool, moist, glacial-relict, or other similar microclimates prevail.

Range: Natural occurrence limited to a few counties located primarily along a NW–SE axis across Indiana from Fountain County to Jefferson County.

Comments: Light, soft, weak, brittle wood is light-red brown. None harvested for timber in Indiana. Bark an important source of tannin. Widely planted as an ornamental evergreen.

TAMARACK

Larix laricina (Du Roi) K. Koch

Distinguishing Features: Small to medium-sized trees (to 70 feet tall, 1–2 feet diam.) with an open, pyramid-shaped, irregular crown of horizontal branches. **Pale green needles are borne many to a cluster. Woody cones are oblong (to 1 1/2 inches).**

Leaves: **Needles are numerous in soft, light-green clusters borne on woody spurs on twigs;** they are about 1-inch long, triangular in cross section. Needles turn brilliant yellow in autumn, then drop—a deciduous "evergreen."

Bark: Thin, reddish-brown, peeling, and scaly.

Twigs/Buds: Twigs slender, light orange-brown, smooth; leaf scars alternate, elevated, borne on spurs. Buds round, conspicuous, but small (1/8 inch diam.), reddish-brown.

Flowers/Fruit/Seeds: Cones unisexual on same tree; male—round, yellow, near branch tips; female—oblong, rose colored, on older branches near leaves. Woody mature cone oblong (to 1 1/2 inches long) to nearly round, brown. Seeds small (1/8 inch) oval with long wing; wind-scattered.

Habitat: Occurs around natural lakes and in bogs, often in pure stands.

Range: The northern three tiers of counties in Indiana.

Comments: Wood heavy, hard, durable, close-grained; heartwood orange-brown. Used for posts, railroad ties, cabin logs, interior finishing. Sometimes planted as an ornamental. Of no economic value in Indiana. Also called Eastern larch.

EASTERN WHITE PINE

Pinus strobus L.

Distinguishing Features: **Medium to large evergreen trees** (to 120+ feet tall, 2–3 feet diam.); its pyramid-shaped crown becomes flat-topped and irregular with age. **Needles soft, blue-green, in clusters of 5. Woody cones the shape of human index finger, only longer.**

Leaves: Blue-green needles in clusters of 5, long (to 5 inches), slender, flexible, soft to the touch, fragrant and resinous when broken.

Bark: Thin, gray, and smooth on young trees; becoming nearly black, deeply furrowed with broad ridges composed of purple-tinged scales when mature.

Twigs/Buds: Twigs slender, smooth, green early, then brown-orange. Buds red-brown, oval, pointed (to 1/2 inch long).

Flowers/Fruit/Seeds: "Flowers" unisexual on same tree; male cones numerous, small, yellow spikes (to 1/3 inch) on new twig growth; female cones fewer, grouped, red to purple. **Mature cones woody, curved (to 8 inches long), frosted white with dried resin,** mature in two seasons. Seeds to 1/4 inch, with wing to 3/4 inch; wind-scattered.

Habitat: Moist woods and wooded slopes on well-drained soils.

Range: Native in only a few counties in northwestern Indiana, plus along certain creeks in west-central Indiana. Now planted ornamentally and for forestry purposes throughout the state. Sometimes regenerates from plantings.

Comments: Wood lightweight, soft, straight-grained, nearly white to light brown. Once used for ship masts, window sashes and doors, house construction, and furniture. Now used for cabinets, house interiors, and carving blocks.

JACK-PINE

Pinus banksiana Lambert.

Distinguishing Features: **Small to medium evergreen trees** (to 50 feet tall, 1+ feet diam.), crown varies from open and cone-shaped to scrubby and distorted. **Needles short (to 1 1/2 inches), stiff, in clusters of 2. Cones woody, curved, frequently closed at maturity.**

Leaves: Needles in clusters of 2, stout, stiff, sharp-pointed, often twisted, dark green; to 1 1/2 inches long.

Bark: Dark red-brown with shallow, rounded ridges; rough and scaly.

Twigs/Buds: **Twigs** yellow-green, then **purplish-brown with age and roughened by leaf bases which remain.** End bud 1/4-inch long, oval, rounded tip, pale brown. Side buds similar but smaller.

Flowers/Fruit/Seeds: "Flowers" unisexual on same tree; male cones small (to 1/3 inch), yellow, at branch tips; female cones clustered, round, dark purple, near branch tips. **Mature cones** woody (1 1/2–2 inches long), **scales without prickles, curved toward branches; often remain closed on tree for several years.** Seeds triangular, black, winged; wind-scattered, usually after fire.

Habitat: Sandy, infertile soils; dry wooded slopes; edges of interdunal ponds.

Range: Natural range in Indiana is a narrow band in dunes area of the four northwesternmost counties. Now widely planted in Indiana, especially on old strip-mined spoil banks.

Comments: Wood moderately hard, close-grained, weak, light reddish-brown. Used for railroad ties, posts, pulpwood. Of little economic importance in Indiana. Seed germinates following fires, often producing dense thickets of small, stunted trees.

VIRGINIA-PINE; SCRUB-PINE

Pinus virginiana Miller.

Distinguishing Features: **Small to medium-sized evergreen trees** (to 50 feet tall, 1–2 feet diam.) with an open, rounded to flat-topped crown, and short trunk. **Needles in bundles of 2 are 1 1/2–3 inches long. Cones woody, 1 1/2–3 inches long, remain on tree, but do open.**

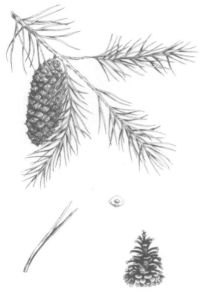

Leaves: **Needles** gray-green, in clusters of two (1 1/2–3 inches long), shiny, stout but flexible, tips diverge, **usually twisted,** tiny teeth on edges, fragrant when crushed.

Bark: Thin, dark brown, scaly; shallowly furrowed between flat plates.

Twigs/Buds: Twigs slender, tough; green early, then gray-brown or purplish at maturity.

Flowers/Fruit/Seeds: "Flowers" unisexual on same tree; male cones small (to 1/3 inch), orange-brown, crowded near branch tips; female cones larger, rounded, light green, on older twigs. Mature cones woody, reddish-brown (1 1/2–3 inches long); **scales armed with a 1/8-inch curved spine; cones open, but remain on tree for a few years.** Seeds pale brown (1/4 inch), winged; wind-scattered.

Habitat: Crests of ridges (usually sandstone), or dry upper wooded slopes.

Range: Natural distribution was a few counties along the Knobstone Escarpment in far southern Indiana. Now somewhat widely planted, especially on old strip-mined sites.

Comments: Wood is light, soft, and brittle—pale orange color. Used mainly for pulpwood, rough lumber, or firewood. Economic value not large in Indiana. Easily confused with Jack-pine, which has shorter leaves and no spine on cone scales.

BALD CYPRESS

Taxodium distichum (L.) Rich.

Distinguishing Features: Large tree (to over 100 feet tall, 4+ feet diam.) with an open crown, pyramid-shaped when young, widely spreading when old. **Trunk often swollen at base. Needle leaves borne singly are pointed at tip. Cones are spherical, about 1 inch in diameter. Has vertical, woody root growths called knees.**

Leaves: Needles borne singly (to 3/4 inch long), pointed at tip, delicate, feathery, in two ranks along twig, light yellow-green. Needles (with branchlets) turn rusty brown in autumn, then drop off.

Bark: Thin for size of tree; **pale reddish-brown;** shallow furrows divide the broad flat ridges that become fibrous and shreddy with age.

Twigs/Buds: Branchlets slightly drooping, twigs slender, light green early, then reddish-brown by winter; leaf scars absent. Buds small (1/8 inch), rounded, with overlapping scales.

Flowers/Fruit/Seeds: "Flowers" unisexual on same tree; male "cones" in slender, hanging, many-flowered clusters (to 5 inches long); female globe-shaped, green, scaly, at branch tips. **"Fruit" a globe-shaped, wrinkled, woody cone (1 inch diam.);** irregular scales, each containing two winged seeds, compose the cone surface. Seeds typically water dispersed; sometimes scattered by wind or animals.

Habitat: Deep swamps, sloughs, ox-bows, and slough-like creeks, usually with standing water much of the year.

Range: Occurred naturally in only five counties in far southwestern Indiana. Widely planted elsewhere in the state, frequently as an ornamental.

Comments: Wood soft, lightweight, oily, and very resistant to decay. Once used for decking, silos, vats, barrels, posts, railroad ties, and bridges. Of little commercial importance in Indiana today. An extremely long-lived species, with some individuals approaching 1000 years.

NORTHERN WHITE CEDAR

Thuja occidentalis L.

Distinguishing Features: A small to medium **evergreen tree** (to 30 feet tall, 1+ foot diam.), with a dense, conical to pyramid-shaped crown. Bark becomes reddish-brown and shreddy with age. **Branchlets compressed into flattened sprays. Leaves opposite, scale-like, evergreen; occur in pairs, giving a braided appearance; very fragrant when crushed.** Buds tiny, not scaled. "Flowers" tiny, in small unisexual cones. "Fruit" is a small, pale-brown cone (to 1/2 inch long), several scales; seeds tiny (to 1/8 inch), wind-scattered.

Comments: **State Endangered and now occurs at only one site in the Indiana dunes region.** Found in wet sites, especially bog margins. Wood very lightweight (see "Wood Densities"), fragrant, very durable—used in canoe and boat construction or outdoor furniture, where available. A long-lived tree, hence Arborvitae (Tree of life); widely planted as an ornamental.

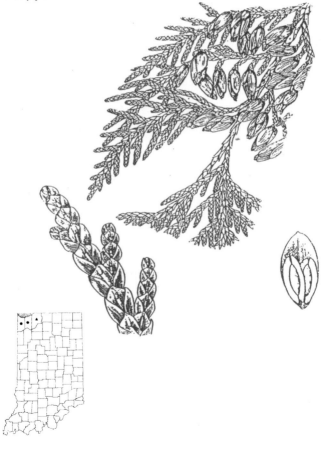

94

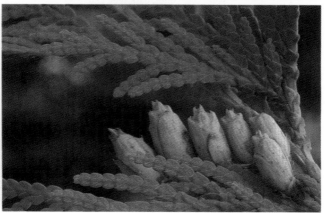

EASTERN RED CEDAR

Juniperus virginiana L.

Distinguishing Features: Small to medium **evergreen tree** (to 50 feet tall, 1–2 feet diam.) with a narrow, cone-shaped, or rounded crown. **Evergreen leaves of two types—(1) sharp, awl-shaped, and (2) scaly, appearing braided. Fruits round, dark blue, berry-like; smell like dry gin.**

Leaves: Evergreen, opposite, of 2 types (both dark green or blue-green): scale-like (to 1/16 inch long), appearing as a braided rope; or needle-like (to 1/2 inch long), sharp, occur singly on twig.

Bark: **Thin, reddish-brown, separating into long shreds.**

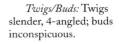

Twigs/Buds: Twigs slender, 4-angled; buds inconspicuous.

Flowers/Fruit/Seeds: "Flowers" unisexual on separate trees; male cone-like, small (to 1/8 inch) at tips of branches; female globe-shaped, small (to 1/10 inch), with fleshy scales. "Fruits" are small (to 1/4 inch), fleshy, berry-like cones; dark blue and fragrant when crushed. Seeds 1–4 per fruit, light brown, ridged; bird-scattered.

Habitat: Abandoned fields, fencerows, dry woods, rocky bluffs and cliffs (especially limestone).

Range: Throughout Indiana, but less common in central tillplain.

Comments: Wood very durable, light in weight, close-grained, rich red color, and fragrant. Used for closet linings, cedar chests, pencils, and fence posts. Cedar oil in wood repels moths. Orange-colored to brown cedar "apples," which are sometimes present, are the fruiting stage of the cedar apple rust fungus.

CUCUMBER-TREE

Magnolia acuminata (L.) L.

Distinguishing Features: Medium-sized tree (to 80 feet tall, 3 feet diam.) with a broad spreading crown. **Large toothless leaves; long buds with a single scale; large showy flowers; cucumber-shaped fruits.**

Leaves: Alternate, simple, borne singly on short (to 1 1/2 inches) stems. Blades oval, large (6–10 inches long, 2/3 as wide), tapering to tip, base rounded; edges smooth to wavy; **upper surface yellow-green and smooth, lower surface pale (nearly white) and hairy.** Turn pale yellow in autumn.

Bark: Thin, dark brown, furrowed with narrow scaly ridges when mature.

Twigs/Buds: Twigs rather stout, reddish-brown to gray, largely hairless; leaf scars U-shaped. Buds large (to 3/4 inch), **silvery-white, hairy, with a single scale.**

Flowers/Fruit/Seeds: Bisexual flowers bell-shaped, greenish-yellow to bluish-green, large (to 3 inches long); 6 petals, odorless; appear in April. **Young fruits look like small cucumbers (hence common name);** mature into red, oblong leathery cones (to 3 inches long) in September/October. Seeds large (to 1/2 inch), scarlet to orange-red; hang from cone on elastic threads when ripe.

Habitat: Rich, moist woodlands.

Range: Now apparently occurs only in Washington County. Possibly in a few counties in south-central and southeastern parts of the state; more widespread in southern Indiana originally. Occasionally planted as an ornamental.

Comments: Wood very similar to Tulip-tree, but heavier. Used for furniture, cabinets, and flooring. Also called Cucumber magnolia, or sometimes Indian bitter.

UMBRELLA-TREE

Magnolia tripetala (L.) L.

Distinguishing Features: Small understory tree (to 40 feet tall, <1 foot diam.) with open spreading crown; often with multiple, twisted, smooth, gray trunks. **Largest simple leaves of any Indiana tree. Flowers very large, white and showy. Large long buds.**

Leaves: Alternate, simple, usually borne clustered near branch tips on short (to 1 1/2 inches) stout leafstalks. Blades very large (to 16 inches long, 1/2 as wide), pointed at tip, tapered at base, edges smooth; papery to leathery texture; bright green above, pale to whitish and hairy below. Turn bright yellow in autumn.

Bark: Thin, light gray, smooth but shallowly furrowed with age.

Twigs/Buds: Twigs stout, smooth, brittle, green to light gray; leaf scars almost circular. **End buds purple, long (1–1 1/2 inches), and pointed.**

Flowers/Fruit/Seeds: Bisexual flowers white, very large (4–6 inches across), showy with 6–9 petals, many reproductive parts. **Fruit leathery and cone-like** (3–6 inches long), green at first, then reddish at maturity. Red seeds (1/4–1/2 inch) suspended from mature fruit by threads.

Habitat: Deep, moist secluded ravines.

Range: Native populations known only from Crawford County.

Comments: Lovely Appalachian species at its northern range limit in Indiana. Wood lightweight, soft, smooth-grained, purplish-gray, satin-textured—of no commercial value. Flowers have strong odor. Also called Umbrella magnolia.

TULIP-TREE

Liriodendron tulipifera L.

Distinguishing Features: Large trees (to 150 feet tall, 4+ feet diam.) with **straight, column-like trunks** and rounded or spreading crowns. **Leaf shape and flowers resemble tulips. Duck-bill shaped buds.**

Leaves: Alternate on twig, simple, borne singly on long (2 1/2–4 inches) slender leafstalks. Blades 4–6 inches across with four broad lobes, the upper two, separated by a broad notch (**"tulip-like" in silhouette**), edges smooth; bright green and smooth above, paler below. Turn bright yellow in autumn.

Bark: Thin, gray, whitish within fissures on young trees; becomes thick, with deep furrows and broad ridges with age.

Twigs/Buds: Twigs moderately stout, smooth, reddish-brown; leaf scars nearly round. Buds long (to 1 inch), green to brown, shiny, with two large scales that resemble a duck's bill.

Flowers/Fruit/Seeds: Bisexual **flowers** borne singly in May; large (to 2 inches); **cup-shaped and tulip-like** (hence name of tree); six green and orange, waxy petals; many reproductive parts. **Fruit an upright cone-like cluster (to 3 inches long) of many seeds.** Winged, angled seeds wind dispersed in autumn; sometimes on snow.

Habitat: Upland woods with rich, moist soil. Often found in fencerows and fertile old fields. A fine ornamental tree and widely planted in lawns and parks.

Range: Throughout Indiana.

Comments: Soft, lightweight (yet strong and durable) wood is yellow-green, fine-grained, and easily worked. Widely used for furniture, cabinetry, interior finish, and construction. Once the prime log cabin timber. Tulip-tree is Indiana's state tree. Called Yellow poplar by foresters.

PAWPAW

Asimina triloba (L.) Dunal.

Distinguishing Features: Trees small (to 30 feet tall, <1 foot diam.). **Leaves very large with smooth edges. Buds, pointed, soft brown, fuzzy. Flowers maroon; fruit large,** green then brown.

Leaves: Alternate, simple, borne singly on very short stems (<1/2 inch); very large (6–12 inches long, 1/2 as wide); widest above middle, pointed tip, narrowing to base, edges smooth; not hairy on surfaces; **emit a distinctive, disagreeable odor when bruised.** Bright yellow in autumn.

Bark: Smooth, gray, often warty but usually without fissures. Sparingly fissured on old trunks.

Twigs/Buds: Twigs slender, rusty brown, fuzzy early, then smooth; leaf scars horseshoe-shaped. **End bud dark rusty brown, long (to 2/3 inch), pointed, flattened,** no bud scales; side buds small, rounded.

Flowers/Fruit/Seeds: Flowers borne singly as leaves unfold in April/May; large (to 1 1/2 inches), brown to maroon, six petals; odor strong, rancid. **Fruit a large (3–5 inches) oblong green berry;** turns brown to black when ripe in October; each contains several large (to 3/4 inch) glossy brown seeds. Fruit pulp edible, nutritious—often called "Indiana banana."

Habitat: Woods and thickets, especially on rich, moist soils along streams; often forms large colonies—"pawpaw patches."

Range: Throughout Indiana.

Comments: Wood weak, soft—used by pioneers for carving wooden spoons, ladles, and fishing bobbers. Could be domesticated for fruit production; many consider it to be delicious when fully ripe.

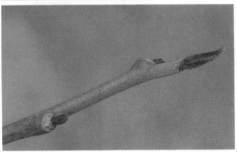

SASSAFRAS

Sassafras albidum (Nutt.) Nees.

Distinguishing Features: Medium-sized trees (to 80 feet tall, 2+ feet diam.) with spreading crowns; **branches often crooked, sometimes twisted in appearance. Variable leaves of oval, single-mitten, and double-mitten shapes. Green twigs and buds. Red-brown bark. Spicy aroma of bark, leaves, twigs, and roots.**

Leaves: Alternate, simple, borne singly on short stems (to 1 1/2 inches); blades (2–6 inches long) of three different shapes— 3-lobed, 2-lobed, or oval and unlobed, often on same tree; **leaf edges smooth, veins conspicuous; spicy odor when crushed.** Turn brilliant orange and red in autumn.

Bark: Green on young trees and branches; becomes deeply fissured and a rich reddish-brown when mature.

Twigs/Buds: Twigs slender, green, smooth, fragrant. Greenish buds egg-shaped, to 1/4-inch long, dull-pointed at tip.

Flowers/Fruit/Seeds: Male and female flowers borne on separate trees before leafing; both yellow-green in clusters. **Fruits are egg-shaped blue-black "berries" in clusters, seated on fleshy, bright red stalks.** Ripen August/September. Largely bird- and small mammal–scattered.

Habitat: Roadsides, fencerows, old fields, upland woods. Often root sprouts into large colonies.

Range: Throughout Indiana.

Comments: Widely known as source of sassafras tea, which is brewed from inner bark of freshly dug roots, which smell like root beer. Fragrant oil used in soaps, perfumes, and medicines. Wood strong, springy, lightweight, durable in soil. Lumber with its beautiful grain is gaining popularity with woodworkers.

A beautiful tree in all seasons; should be more widely used as an ornamental.

AMERICAN SYCAMORE

Platanus occidentalis L.

Distinguishing Features: Large trees (to 100 feet tall, 6+ feet diam.) with spreading crowns and **white bark on upper trunk and limbs. Leaves large, maple-like. Fruit a "sycamore ball" on drooping stem.**

Leaves: Alternate, simple, borne singly on long (to 4 inches), hairy leafstalks. Blades nearly circular in outline (to 8 inches across), but with 3–5 main sharp-pointed hand-like lobes, the lobes very coarsely toothed; bright green above, paler lower surface with hairy veins. Stipules at leaf bases; leaves turn rusty brown in autumn.

Bark: Brown in thin rounded scales, which flake off to expose **large green or white expanses, especially on upper trunk.**

Twigs/Buds: Twigs moderately stout, zigzag, smooth, green early, brown later; leaf scars circular, surround bud. **Buds cone-shaped, 1/4-inch long with a single scale; each bud covered completely in summer by enlarged, hollow leaf stem base.**

Flowers/Fruit/Seeds: Flowers unisexual on same tree, but borne separately in May; crowded together in dense, round clusters. Fruits are round, light brown, 1-inch diameter clusters of many seeds, borne on long, drooping stems. Fruits ripen in October; often retained throughout winter. Each seed has a plume of hairs for wind dispersal.

Habitat: Occurs widely on rich bottomlands of creeks and rivers, around lakes and ponds, and invades old fields. Frequent ornamental or shade tree, but messy due to shedding of bark, branches, leaves, and fruits.

Range: Throughout Indiana on appropriate sites.

Comments: Wood hard, tough, and has a pretty grain, but lumber often warps. Being almost unsplittable, it is used for butcher blocks and crating; occasionally for furniture and paneling; has beautiful grain when quarter-sawed. Mature trees often have anthracnose blight (a fungal disease), which blackens leaves, especially in wet seasons. Also called Buttonwood from button-like fruit clusters.

SWEET GUM; RED GUM

Liquidambar styraciflua L.

Distinguishing Features: Large trees (>100 feet tall, 4 feet diam. on prime sites), with spreading crowns and resinous sap. **Star-shaped, alternate leaves and spiny "gum ball" fruits. Corky ridges on twigs.**

Leaves: Alternate on twig, simple, borne singly on leafstalks 2–3 inches long. Blades 2–5 inches across; star-shaped with 5 (occasionally 7) points; toothed edges; smooth, lustrous green above, paler and slightly hairy below. Their bright autumn colors range from yellow to red to purple.

Bark: Pale gray and smooth on young trees and upper branches. Dark gray and deeply fissured into rough scaly ridges when mature.

Twigs/Buds: Twigs stout, green-brown early, gray and often with corky wings at maturity. Leaf scars half round with three bundle scars. **Buds large, shiny, pointed with many green to brown scales; often sticky to touch.**

Flowers/Fruit/Seeds: Flowers unisexual on same tree in May. Male flowers in large (2–3 inches) green clusters; female flowers clustered on swinging (2–3 inches) stems, the globe-shaped heads later mature into round, clustered, **spiny "gum ball" fruits (1–2 inches diameter).** Many tiny dark brown seeds are wind-scattered.

Habitat: Wet bottomlands and flatwoods. Widely planted as ornamental shade tree of high quality.

Range: Most of Indiana south of Indianapolis.

Comments: Excellent timber tree. Wood takes a high polish—used for furniture, interior trim, and cabinetry. Hard, strong lumber sold as red gum, somewhat resembles walnut. Frequently grows with pin oaks in natural plant communities.

Generic name means "fluid amber" from its resin, which is used as storax, a fragrant balsam used in medicines and perfumes.

AMERICAN ELM

Ulmus americana L.

Distinguishing Features: A handsome medium to large tree (to 100 feet tall, 4 feet diam.) with a broad, rounded crown. Open-grown trees have diverging limbs and drooping branches, resulting in a **vase-shaped growth form. Bark has alternating brown and white layers.**

Leaves: Alternate on twig, borne singly on very short (<1/2 inch) yellow stems. Blades to 5 inches, 1/2 as wide; **edges doubly saw-toothed; upper surface dark green and smooth to touch,** lower surface paler and hairy to smooth, leaf bases asymmetrical. Turn yellow in autumn.

Bark: Medium to dark gray; older trees have deep, irregular, intersecting furrows separated by flat ridges. **A cross-section of a piece of bark reveals alternating layers of nearly white cork and dark reddish-brown fiber.**

Twigs/Buds: Twigs slender, smooth to hairy, red-brown to gray. Leaf scars alternate, half-round. Buds to 1/4 inch, pointed, with small, red-brown, hairy scales.

Flowers/Fruit/Seeds: Bisexual flowers small, clustered on slender, drooping stalks before leafing; greenish-red, hairy, small. **Fruits dry, in dense clusters, flattened, oval (<1/2 inch); a circular, papery wing surrounds seeds; wing hairy-edged and notched at top;** wind-scattered in spring.

Habitat: Most common in bottomland forests along streams and on floodplains. Also occurs on upland flats and in moist ravine forests, in fencerows, and old fields.

Range: Throughout Indiana.

Comments: Pale brown wood is heavy, tough, and springy; warps badly. Used for boxes, crates, barrels, and steam-bent wooden items. Once the most popular shade and ornamental tree in eastern United States. Now widely killed by Dutch elm disease and phloem necrosis. Commonly called White elm.

SLIPPERY ELM

Ulmus rubra Muhl.

Distinguishing Features: Medium to large trees (to 100 feet tall, 3+ feet diam.) with broadly rounded or flat-topped open crown. **The only elm with leaves having asymmetrical bases and rough, sandpapery upper surface. Bark uniformly reddish-tan in cross-section.**

Leaves: Alternate on twig, borne singly on very short (<1/2 inch), stout, hairy stems. Blades to 6 inches long, 1/2 as wide; bases strongly asymmetrical, tips pointed; edges coarsely doubly saw-toothed. Dark green above and very rough to the touch; paler and less rough below.

Bark: Gray-brown to reddish-brown with shallow furrows and flattened ridges; outer bark uniformly reddish-tan to light brown in cross-section. **Inner bark with a sticky, slippery sap** (hence common name). Sap once used in medicines to ease sore throats.

Twigs/Buds: Twigs moderately stout, gray to red-brown, hairy, rough to the touch. Leaf scars alternate, half-round. Buds oval (to 1/4 inch), dull-pointed; covered with rusty-colored, hairy scales.

Flowers/Fruit/Seeds: Bisexual flowers in dense greenish to reddish clusters on very short stalks before leaves emerge. **Fruits dry, with papery, circular wing** (to 3/4-inch across) surrounding a single seed. **Wing edges not hairy; shallowly notched at top.** Seeds wind-scattered April–May.

Habitat: Floodplain forests and bottomland woods. Also found in mesic (sometimes even rocky) upland woods.

Range: Throughout Indiana.

Comments: Dark brown wood heavy, strong, tough; its pretty grain resembles oak. Once used in shipbuilding and wagon manufacture; now for crates, boxes, some furniture. Little used ornamentally. Less susceptible to disease than American elm. Also called Red elm, especially the lumber.

ROCK-ELM; CORK ELM

Ulmus thomasii Sarg.

Distinguishing Features: Medium tree (to 80 feet tall, 3 feet diam.) with a narrow, oblong crown of drooping branches. **Twigs usually with corky ridges or wings. Leaves alternate, simple on smooth leaf stalk.**

Leaves: Alternate on twig, borne singly on very short (<1/2 inch), smooth stems. Blades oval (to 4 inches long, 1/3 as wide), edges doubly saw-toothed; short-pointed tip, tapering to even base. Surface green, shiny, and smooth above, paler and hairy below; somewhat rough to the touch. Turn yellow in autumn.

Bark: Thick, grayish with wide fissures separating broad, flat, scaly ridges. Alternating brown and white layers in cross-section; **often very corky.**

Twigs/Buds: **Twigs** slender, brownish, smooth, **often with corky, wing-like ridges.** Leaf scars alternate, half-round. Buds small (to 1/4 inch), slender, pointed, with brown, somewhat hairy scales.

Flowers/Fruit/Seeds: Bisexual flowers occur along central twig axis before leafing in slender, drooping, branched clusters of 2–4 individual flowers. **Fruits dry,** oval (to 1/2 inch long), hairy; **a papery wing surrounds the single seed; tip of wing notched and edge of wing hairy.** Fruits, in season, eaten avidly by squirrels. Seeds usually wind-scattered in April–May.

Habitat: Upland forests, especially on slopes, rocky ridges (hence common name), and river bluffs. Also found on rich, moist soils of beech-maple woods.

Range: Scattered, largely in northeastern two-thirds of state.

Comments: Pale brown wood is hard, strong, heavy, and close-grained. Used formerly for axe handles, hockey sticks, and farm implements. Easily split into fence rails by the pioneers, who usually referred to Rock-elm as "hickory elm." Also called Cork elm.

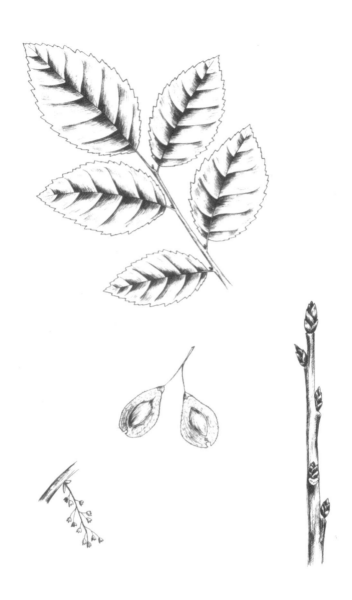

WINGED ELM

Ulmus alata Michx.

Distinguishing Features: **Small to medium tree** (to 50 feet tall, 1–2 feet diam.) with narrow, oblong crown. **The combination of corky wings on smaller branches, short, hairy leaf stems and small leaves separate it from all other elms.**

Leaves: Alternate on twig, borne singly on very short (<1/4 inch), stout, hairy leafstalks. Blades to 3 inches long, 1/2 as wide are doubly saw-toothed on edges. More slender than other elms. Dark green and smooth above, paler and downy below. Turn yellow in autumn.

Bark: Gray to brown with shallow furrows separating flat ridges.

Twigs/Buds: **Twigs** slender, light green then gray to red-brown; usually **with very pronounced thin, corky ridges or wings** (hence common name). Leaf scars alternate, half round. Buds small (to 1/4 inch), slender, long-pointed, sometimes curved, with tiny, overlapping, rusty-brown scales.

Flowers/Fruit/Seeds: Bisexual flowers occur before leafing, in short drooping clusters of 3 to 7; bell-shaped, green to reddish. **Fruits dry, oval (to 1/3 inch long) with a papery wing around a single, hairy seed; wing tip has small notch.** Seeds wind-scattered in spring.

Habitat: Dry ridges, bluffs, old fields; occasionally in low woods.

Range: **Confined to southwestern fifth of Indiana.**

Comments: Wood heavy, hard, brittle, pale brown, but primarily an understory or small tree of little commercial value. Used for tool handles, small articles. Occasionally planted as an ornamental.

SOUTHERN HACKBERRY; SUGARBERRY

Celtis laevigata Willd.

Distinguishing Features: Medium to large tree (to 80 feet tall, 2 1/2 feet diam.) with broad open crown of drooping branches. Leaves alternate, simple, lance-shaped. **Gray, warty bark. Fruit a small "berry" with hard seed.**

Leaves: Alternate on twig, borne singly on short (to 1/2 inch) leafstalks. **Blades (to 6 inches long) lance-shaped, long tapering tips, base unequal, edges mostly without teeth,** surfaces smooth to slightly rough.

Bark: Gray-brown to silver gray and smooth on young trees and upper branches; older trees have smooth, gray bark with warty projections, but not rough and ridged like hackberry.

Twigs/Buds: Twigs very slender, smooth, zigzag; greenish early, then red-brown. Leaf scars alternate, small, crescent-shaped. Triangular buds small (to 1/8 inch), pointed, hairy scales dark brown.

Flowers/Fruit/Seeds: Male, female, and bisexual flowers on same tree; male in clusters, others solitary; greenish-yellow, without petals. **Fruits fleshy, round (to 1/3 inch), orange-red to yellow "berries" on a long stalk;** with a single hard seed. Fruit sweet (hence name of tree), edible; will float, so frequently water-scattered.

Habitat: Swampy woods of low-lying sites, streamsides, and frequently flooded bottomlands. Occasional on moist slopes or uplands.

Range: Largely in southwestern Indiana. Scattered in wet woods elsewhere in southern Indiana.

Comments: Heavy, close-grained wood is pale yellow. Low economic value; uses similar to Northern hackberry.

NORTHERN HACKBERRY; HACKBERRY

Celtis occidentalis L.

Distinguishing Features: Medium to large tree (to 90 feet tall, 3 feet diam.) with alternate, simple leaves. **Open, spreading, twiggy crown, often with brushy "witches' brooms." Gray bark with distinctive warty bumps. Fruit a small brown-purple "berry."**

Leaves: Alternate on twigs, borne singly on short (to 1 inch) leafstalks. **Blades (to 6 inches long) long, pointed, base uneven, edges toothed,** smooth to roughened above, hairy below along the prominent veins. Leaf surfaces often have fleshy galls.

Bark: Gray to brown and smooth on young trees; soon covered with "warty" bumps. Mature trees with dark gray roughened ridges.

Twigs/Buds: Twigs slender, zigzagged, gray to light brown, smooth; pith somewhat chambered. Leaf scars alternate, small, crescent-shaped. Buds small (to 1/4 inch), pointed, scales with fine hairs; **bud tip pressed closely to twig.**

Flowers/Fruit/Seeds: Unisexual or bisexual on same tree; green, without petals; female flowers singly, male in clusters. Fruits fleshy, round (to 1/3 inch); borne on long stalks; brown-purple when mature; flesh orange-brown, sweet, edible. One large (to 1/4 inch), hard, round seed.

Habitat: Usually occurs along streams, on floodplains, or in low wet woods. Occasional in fencerows or moist upland woods. Use as an ornamental is increasing.

Range: Throughout Indiana.

Comments: Subject to infestation by gall-producing insects. Crowns often have twiggy "witches' brooms." Wood heavy, pale yellow, close-grained. Used for fuel, inexpensive furniture, boxes, crates, and pallets.

OSAGE-ORANGE

Maclura pomifera (Raf.) C. K. Schneider.

Distinguishing Features: Short, medium-sized tree (to 40 feet tall, 1 1/2 feet diam.) with rounded growth form; **twisted branches have spines. Wood orange; sap milky. Leaves dark, glossy green; pointed tips; edges smooth. Very large, round, yellow-green fruits.**

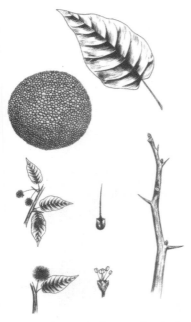

Leaves: Alternate on twig, simple, borne singly on short (to 1 1/2 inches) leafstalks. Blades oval (3–5 inches long, 2/3 as wide) with long tapering tips; edges smooth; dark green and shiny above, paler and somewhat hairy below. Leaf stems emit a sticky, milky sap when broken.

Bark: Gray-brown tinged with orange, fissured into shaggy strips. **Root bark bright orange.**

Twigs/Buds: **Twigs** slender, dull orange-brown, zigzag; armed **with short, sharp spine at each leaf stem** ("nature's barbed wire"). Buds very small, round, reddish-brown.

Flowers/Fruit/Seeds: Flowers unisexual on separate trees. Male flowers crowded in clusters on long (to 4 inches) stems; female flowers in round many-flowered heads on short, stout stalks. Fruit large, green to yellow, rough, multiple cluster of many hard, fused seeds—to size of grapefruit—and filled with milky sap.

Habitat: Hedgerows, old pasture fields, occasionally in low-ground woods.

Range: Native of southwest Arkansas and adjacent parts of Oklahoma, Louisiana, and eastern Texas. Introduced into Indiana a century ago as a living hedge fence and windbreak, especially in prairie areas. Escaped and occurs widely in Indiana now.

Comments: Very dense, hard, flexible, durable wood. Excellent for fence posts. Used by Native Americans as a bow wood, and by the pioneers for wagon wheel hubs. Fruit called a hedge-apple—also another name for the tree.

RED MULBERRY

Morus rubra L.

Distinguishing Features: Small to medium-sized tree (to 50 feet tall, 1+ feet diam.), crown broadly rounded. **Leaves variable in shape with toothed edges. Fruit a distinctive multiple long red berry. Root bark is yellow.**

Leaves: Alternate, simple, borne singly on leafstalks to 1 1/2 inches. Blades 3–5 inches long, nearly as wide; edges coarsely toothed; shape variable—heart-shaped, single mitten, double mitten. Dark blue-green and rough above, paler and hairy below; turn bright pale-yellow in autumn.

Bark: Thin, dark brown; divided into long scaly plates on mature trees.

Twigs/Buds: **Twigs** slender, greenish to brown, sometimes hairy; **exude milky sap when broken.** True end bud absent; side buds oval, pointed, small (to 1/4 inch) with several green to brown scales.

Flowers/Fruit/Seeds: Flowers unisexual, sexes on same or separate trees, clustered in hanging spikes; flower as leaves unfold in May. Fruit multiple (forms from several individual flowers), green to red to purple to black as they mature in June/July; long (to 1 1/2 inches) and drooping. Edible, sweet, and juicy—loved by birds, squirrels, and children. A tiny hard seed in each section of berry. Bird-scattered.

Habitat: Moist woods, especially along streams. Fencerows, old fields, and waste areas. Common in cities although seldom planted there.

Range: Throughout Indiana.

Comments: Wood hard, heavy, yellow-brown, very durable in soil—excellent for fence posts. Little used commercially. Leaves not suitable to silkworm production. Fruit excellent for pies.

BUTTERNUT

Juglans cinerea L.

Distinguishing Features: Small to medium tree (to 75 feet tall, 1–2 feet diam.) with straight, clear trunk and broad, open crown. **Leaves large, compound, alternate. Twigs with chambered pith. Fruits are large, oval nuts with husks.**

Leaves: Alternate on twig, large (to 2 feet long), singly compound with 11 to 17 ladder-like leaflets, **one stemmed leaflet at leaf tip.** Leaflets 2–4 inches long, finely saw-toothed, yellow-green, rough above, often sticky, hairy below. Yellow-brown in autumn; drop early.

Bark: **Divided in broad, whitish (hence white walnut) ridges** with rather shallow furrows; often in roughly a diamond-shaped pattern.

Twigs/Buds: Stout, green to brown, leaf scars shield-shaped with hairy pads ("monkey-face") at top. **Pith chocolate brown, chambered with thick partitions.** Buds cone-shaped; end bud (to 3/4 inch long), blunt tip, hairy; side buds smaller.

Flowers/Fruit/Seeds: Flowers unisexual on same tree. Male flowers in long (2–5 inches), yellow-green, drooping clusters (catkins); small female flowers at twig ends. **Fruits large** (2–3 inches long), **oblong,** usually in pairs or clusters, green, becoming brown at maturity; **sticky, hairy, thin husk does not open.** Large rough seeds usually four-angled. Nuts oily, edible, buttery; delicious at harvest, rancid later. Scattered by squirrels.

Habitat: Most common on well-drained floodplains and creek bottoms. Occasionally as scattered individuals or small groves in mixed woods.

Range: Once widespread in most of Indiana. Now uncommonly to rarely found.

Comments: Wood light brown, rather soft with definite grain. Once used in furniture and cabinets. Trees now susceptible to disease and populations are declining. Called "white walnut" by pioneers and many lumbermen. Husks of nuts were the source of the dye used to color the uniforms of soldiers of the Confederate Army during the American Civil War— hence the soldiers were called "butternuts."

Juglandaceae 129

BLACK WALNUT

Juglans nigra L.

Distinguishing Features: Large tree (to 120 feet tall, 3+ feet diam.), often with straight trunk for half its height; crown broad and open. **Twigs stout, brown; pith chambered. Leaves large, alternate, compound. Fruit a large round nut with soft husk. Inner bark is rich chocolate brown.**

Leaves: Alternate on twig, large (to 2 feet long), singly compound with 15–23 ladder-like leaflets, each 3–4 inches long, finely saw-toothed, yellow-green, smooth above, hairy below. **Usually with two leaflets at tip of leaf.** Leaves emerge late in spring; they turn yellow to brown and drop early in autumn.

Bark: On young trees scaly and gray to brown. Mature trees have thick, dark brown to black bark, furrowed to blocky with broad ridges. **Inner bark a rich chocolate brown** in criss-crossing layers.

Twigs/Buds: Twigs stout, light brown; **tan pith is chambered with thin partitions.** Leaf scars alternate, elevated, shield-shaped. End buds brown, large (to 1/2 inch), blunt, hairy; side buds smaller.

Flowers/Fruit/Seeds: Flowers unisexual on same tree. Male flowers in long (2–4 inches) yellow-green, hairy, drooping clusters (catkins); female flowers small on short spikes at twig ends. **Fruits large (1 1/2–3 inches diam.) round nuts in pairs, with soft green to yellow husk.** Deeply ridged nut surface. Nuts edible, delicious. Scattered by squirrels.

Habitat: Prefers deep, well-drained fertile soils of infrequently flooded floodplains or moist upland woods. Occurs as scattered clumps or individuals.

Range: Throughout Indiana.

Comments: Wood is heavy, hard, chocolate brown with beautiful grain. Prized for furniture, cabinets, gunstocks, and woodworking. Tree contains juglone, a chemical that is toxic to many herbaceous plants.

PECAN

Carya illinoensis (Wangenh.) K. Koch.

Distinguishing Features: Large trees, largest of the hickories (to 150+ feet tall, 4+ feet diam.); crowns narrow at base, spreading at top; trunk straight and clear. **Leaves compound, with up to 17 leaflets. Fruits elongated, with thin husks and thin shells. Buds long with 2 hairy scales.**

Leaves: Alternate on twig (to 20 inches long), singly compound with 9 to 17 ladder-like, lance-shaped (often curved) leaflets, tapering to pointed tips. Leaflets bright green to yellow-green above, paler below, hairy to smooth. Turn yellow to brown in autumn.

Bark: Thick, light brown when young to reddish-brown at maturity. With deep narrow furrows alternating with rough ridges.

Twigs/Buds: Twigs short, stout, light brown with orange pores, hairy. Leaf scars large, 3-lobed. End buds medium (to 1/2 inch), flattened with two yellow-brown, hairy scales.

Flowers/Fruit/Seeds: Flowers unisexual on same tree. Male flowers in slender, long (4–6 inches), stalked, hairy clusters (catkins); female in few-flowered spikes at twig tips. **Fruits oblong (1 1/2–2 inches),** borne in clusters of 3–6; **husk thin, splitting to base at maturity.** Nuts smooth, brown to reddish, thin-shelled; kernels sweet, oily, delicious.

Habitat: Fertile, moist soils of the bottom-lands along larger rivers and streams.

Range: **Southwestern Indiana along major watercourses,** especially the lower Wabash, White, and Ohio Rivers and their tributaries.

Comments: Widely planted and cultivated for their nuts. Nuts important to wildlife. Pale red-brown wood is used for furniture, cabinets, paneling, and tool handles.

BITTERNUT-HICKORY

Carya cordiformis (Wangenh.) K. Koch.

Distinguishing Features: Medium to large trees (to 90 feet tall, 3 feet diam.), with straight trunks and narrow crowns, but spreading with age. **Leaves compound with 7–9 leaflets. Bark tight, not scaly or shaggy. Buds bright sulfur yellow. Nuts small with thin husks and shells. Kernels very bitter.**

Leaves: Alternate on twig (to 1 foot long), singly compound with 7–9 ladderlike, narrow leaflets, each tapering to a narrow point, and with saw-toothed edges. Bright green above; paler, often hairy below. Turn bright yellow to brown in autumn.

Bark: Shallow fissures and flat ridges are interlaced, but remain tight on trunk and do not become scaly or shaggy.

Twigs/Buds: **Twigs slender,** green to gray-brown; leaf scars small, oval. **End buds** flattened, medium (to 3/4 inch long), usually **with** only **two bright sulfur yellow, hairy scales that do not overlap.**

Flowers/Fruit/Seeds: Flowers unisexual on same tree. Male flowers in slender clusters of medium length (3–4 inches); female flowers small, on short spikes. **Fruits small (to 1 inch); husk yellowish, thin (<1/8 inch), leathery, splits only halfway to base, winged above middle.** Nut smooth, thin-shelled, very bitter.

Habitat: Occupies a variety of sites from rich bottomlands where it does best to moist upland woods to drier hillsides. Widespread.

Range: Throughout Indiana.

Comments: Bitternut-hickory is more similar to pecan than to other hickory species. Wood has high shock resistance and high fuel value; its uses are similar to other hickories. Sometimes called Yellowbud hickory.

SHELLBARK-HICKORY

Carya laciniosa (Michx. f.) Loudon.

Distinguishing Features: Large trees (to 110 feet tall, 2–3 feet diam.), very **similar in appearance to Shagbark-hickory** (which see), except leaflets of this species are usually 7 (5–9), and the more-**elongated nuts are much larger (to 2 1/2 inches across).**

Leaves: Alternate on twig (to 1 1/2 feet long), singly compound with 5–9 (typically 7) ladder-like leaflets; the upper three leaflets much larger than the others. Yellow-green, hairy in summer; yellow to brown in autumn.

Bark: Medium thickness, light gray; forms long, very hard plates with ends curving away from the trunk, giving a shaggy appearance.

Twigs/Buds: **Twigs stout, orange-brown; hairy when young,** then smooth, with numerous orange pores. Leaf scars alternate, 3-lobed. End buds large (to 1 inch), oval, with several dark-brown overlapping scales; side buds much smaller.

Flowers/Fruit/Seeds: Flowers unisexual on same tree. Male flowers in very long (4–7 inches), yellow-green slender drooping clusters; small, densely hairy female flowers on short spikes at end of twig. **Fruits very large, oblong to round (to 2 inches by 2 1/2 inches long), with thick (to 3/4 inch) woody husk that splits completely at maturity.** Nuts pale brown, with very hard, thick shells. Kernels sweet, edible.

Habitat: Typically on deep, fertile, moist soils of floodplains and infrequently flooded low ground.

Range: Most of the state except the prairie border region of northwestern Indiana.

Comments: Wood similar to Shagbark-hickory and used for the same purposes. All hickory lumber is sold as one kind. Nuts are an important food source for forest animals. Also called Kingnut hickory.

MOCKERNUT-HICKORY

Carya tomentosa (Poiret) Nutt.

Distinguishing Features: Medium to large tree (to 100 feet tall, 2–3 feet diam.); **similar to shagbark hickory**, except the **bark is much less shaggy, and usually has 7 leaflets, which are hairy below.**

Leaves: Alternate on twig (to 1 foot long), singly compound with 5–9 (usually 7) ladder-like leaflets; the upper three leaflets larger than the others; darker yellow-green above, hairy below. Turn bright yellow to brown in autumn. May retain leaves longer than other hickories.

Bark: Thick, dark gray with shallow irregular furrows and hard narrow flat ridges in a criss-crossing pattern. Somewhat ash-like. Normally does not separate into long loose strips.

Twigs/Buds: **Twigs stout, hairy,** gray to red-brown, pores conspicuous; leaf scars large, 3-lobed. **End buds large (to 3/4 inch), rounded to oval, with 3–4 overlapping, dark, hairy scales;** side buds smaller.

Flowers/Fruit/Seeds: Flowers unisexual on same tree. Male flowers in long (4–6 inches), light green, hairy, slender, drooping clusters; small female flowers are hairy, on short spikes at twig ends. **Fruits medium to large (1 1/2–2 inches), with medium (to 3/8 inch) woody husk that splits completely at maturity.** Nuts reddish-brown, pointed ends, thick shells usually 4-angled. Kernel sweet, tasty.

Habitat: Mainly in drier forests along ridges and hillsides on fertile, well-drained soils, but may occur in southwestern Indiana lowlands.

Range: Primarily in hilly landscapes of southern and northern Indiana. Less often found on central tillplain.

Comments: Wood similar to Shagbark-hickory; sold and used together as one type of lumber. Nuts have thicker shells than shagbark and kernels are harder to extract, hence "mockernut." Also called White hickory.

SHAGBARK-HICKORY

Carya ovata (Miller.) K. Koch.

Distinguishing Features: Medium to large tree (to 100 feet tall, 2–3 feet diam.) with narrow (often rectangular) open crown of large branches; **trunk straight with shaggy bark. Leaves alternate, compound. Fruit a large nut with a thick woody husk.**

Leaves: Alternate on twig (to 1 foot long), singly compound with 5–7 smooth (typically 5), ladder-like leaflets with fine saw-toothed edges. Terminal three leaflets much larger than others. Bright yellow to brown in autumn.

Bark: Gray, thin to medium thickness; on mature trees bark breaks into very hard thin plates which are free from the trunk at one or both ends, giving a shaggy appearance.

Twigs/Buds: **Twigs stout,** gray to red-brown, hairy when young; darker and smooth with age. Leaf scars triangular. End buds large (to 3/4 inch), oval, blunt at tip, with 3–4 overlapping, hairy, loose scales.

Flowers/Fruit/Seeds:
Flowers unisexual on same tree. Male flowers on long (4–5 inches), light green, slender, drooping clusters (catkins); small female flowers on short spikes at twig ends. **Fruits large (to 1 1/2 inches diam.), globe-shaped nuts with woody husk to 1/2 inch thick. Husk separates completely into four parts at maturity.** Nuts pale-colored; very hard shell is 4-angled. Kernels sweet, edible, delicious.

Habitat: In woods mixed with oaks on drier uplands and slopes; occasional on wetter sites.

Range: Throughout Indiana.

Comments: Wood very hard, tough, springy. Used for tool handles, paneling, cabinets, and baskets. Excellent for firewood, charcoal, and smoking meats. Very slow-growing trees. Many hickory species are variable and similar to each other, making identification difficult.

RED-HICKORY

Carya ovalis (Wangenh.) Sarg.

Distinguishing Features: Medium to large trees (to 90 feet tall, 2–3 feet diam.). Quite **similar to pignut-hickory** (which see), but this species **usually has 7 leaflets**; the bark of older trees separates into thick, dark, often scaly ridges; **yellow-dotted buds**; and **fruit husks which split to the base.**

Leaves: Alternate on twig (to 1 foot long), singly compound with 5–7 (usually 7) ladder-like, smooth, toothed leaflets, pointed at tips. Turn bright yellow to brown in autumn.

Bark: Thin, gray, and rather smooth when young. With age it forms deep furrows alternating with broad, dark gray hard scaly ridges.

Twigs/Buds: **Twigs slender,** gray to brown, smooth, pores usually present. Oval end buds medium (to 1/2 inch) with overlapping, usually hairy-edged scales; side buds smaller.

Flowers/Fruit/Seeds: Flowers unisexual on same tree after leaves unfold. Female flowers small, 1–2 on spike near twig tips; male in slender, drooping clusters (to 3 inches long). **Fruits oval (to 1 1/2 inches), nuts with thin (to 3/16 inch), woody husk that splits to the base at maturity.** Nuts somewhat ridged, medium shell; kernel sweet.

Habitat: Usually drier sites on wooded slopes and ridges.

Range: Throughout Indiana, but often not numerous or dominant in a stand.

Comments: Wood darker red-brown than other hickories (hence its name). Wood uses similar to Shagbark-hickory (which see). Also called Sweet pignut-hickory from its tastier fruit. Some authors combine Red and Pignut hickories.

Juglandaceae   143

PIGNUT-HICKORY

Carya glabra (Miller.) Sweet.

Distinguishing Features: Medium to large trees (to 90 feet tall, 2–3 feet diam.), similar overall to shagbark hickory, except the bark is in thin criss-crossing fissures and ridges, and the **fruit husk is thin and splits only part way at maturity.**

Leaves: **Alternate on twig** (to 1 foot long), **singly compound with 5–7 (usually 5) ladder-like, narrow leaflets** with many sharp-pointed teeth. Dark yellow-green above, paler and smooth below. Turn bright yellow to brown in autumn.

Bark: Thin, gray, and rough from numerous interlacing shallow fissures and narrow flattened ridges, but not shaggy.

Twigs/Buds: **Twigs slender,** somewhat angled, gray to red-brown, smooth with numerous pores; leaf scars small, oval. End buds oval, medium size (to 3/8 inch), light brown, smooth, outer scales few to absent.

Flowers/Fruit/Seeds: Flowers unisexual on same tree. Male flowers in slender, hairy clusters (to 3 inches long); female flowers in small clusters at branch tips. **Fruits rounded at top, sharply tapering to narrow base (1–1 1/2 inches across); husk leathery, thin (1/8 inch), not splitting completely at maturity.** Nuts not ridged or weakly ridged, shell medium thickness, kernels sweet to somewhat bitter.

Habitat: Dry woods, hillsides or along ridges in association with Shagbark-hickory and dry-site oaks.

Range: Widespread, but more common in the south half of Indiana.

Comments: Wood light brown, similar in properties and uses to Shagbark-hickory (which see). Nuts an important food source for tree squirrels and other wildlife.

PALE HICKORY

Carya pallida (Ashe) Engler & Graebner.

Distinguishing Features: A small to medium-sized tree (to 50+ feet tall, 2+ feet diam.) with a spreading crown. Leaves alternate on twigs are compound with usually **7 to 9 leaflets, each long and slender with very long, drawn out tips;** edges finely-toothed; shiny green above, very pale and hairy below. Turn yellow to brown in autumn. Buds red-brown with silvery scales. Bark nearly black, deeply fissured. **Fruits small** (to 1 inch), **almost white (hence *pallida*); husk thin, narrowed at base; nut ridged,** nearly round. Wood similar to other hickories, and sold as such, if harvested.

Comments: **Rarely encountered and endangered in Indiana and apparently confined (usually) to fine sandy soils of a few far-southwestern counties,** often in association with Post-oak. Ecological relationships of the species not well known. Similar to Pignut-hickory, except this species has yellow scales near center of leafstalk.

Additional Hickory Species: Black hickory, *Carya texana* Buckley, has also been reported for southwestern Indiana (Knox County area) on dry sandy ridges. Currently it is on the State Endangered Species List. Characteristics include: bark nearly black (hence name); buds densely covered with resinous, yellow scales; leaflets 5–7, petiole and lower surface densely rusty-hairy; fruit, husk thin, splitting to base, nut 4-angled, kernel sweet.

AMERICAN BEECH

Fagus grandifolia Ehrh.

Distinguishing Features: Large trees (to 100 feet tall, 4 feet diam.) with clear, straight trunks and compact rounded crowns of slender, spreading branches. **Bark smooth, pale gray. Fruit a spiny bur containing two pyramid-shaped nuts. Buds long, very sharp-pointed, reddish-brown, scaled.**

Leaves: Alternate on twig, simple, borne on slender, very short, hairy leafstalks. **Blades oval** (to 5 inches long, 1/2 as wide), **leathery, pointed at tip, coarsely single-toothed along edges.** Upper surface blue-green, smooth, shiny; paler and sometimes hairy below. Turn copper-gold in autumn.

Bark: Thin, light gray to blue-gray, smooth; often marred by carvings or initials.

Twigs/Buds: Twigs slender, zigzag, brown to gray. Leaf scars alternate, half round. Buds slender, long-pointed (to 3/4 inch), smooth, with obvious overlapping reddish-brown scales.

Flowers/Fruit/Seeds:
Flowers unisexual on same tree; appear after leafing. Male flowers densely clustered in rounded heads and hanging by a slender stalk (to 2 inches). Female flowers small, paired, on short (to 3/4 inch) stalks at branch tips. **Fruits are 4-parted, spiny burs, each containing two triangular nuts.** Edible and tasty; bird- and mammal-scattered.

Habitat: Most common in rich, moist, upland woods, but also in flatwoods and shallow depressions.

Range: Throughout Indiana, except in few northwestern prairie counties.

Comments: Wood hard, strong, close-grained; light brown to deep red-brown. Used for fuel, flooring, chairs, tool handles, clothespins, and charcoal. Very tolerant of shade and a dense shade producer.

AMERICAN CHESTNUT

Castanea dentata (Marshall) Borkh.

Distinguishing Features: Formerly large trees (to 100 feet tall, 3+ feet diam.) with broadly rounded crowns. **Long spear-shaped leaves have coarsely sharp-toothed edges. Fruit a large spiny bur with 2–3 red-brown edible nuts**.

Leaves: Alternate on twig, simple, borne singly on short (to 1/2 inch), stout leafstalks. Blades are oblong (to 8 inches long, <1/2 as broad), with pointed tips and coarse sharp-toothed edges; shiny, smooth, and yellow-green above, paler and hairy below. Turn yellow-brown in autumn.

Bark: Dark brown, thick, with shallow furrows and broad flat ridges. Inner bark a rich reddish-brown.

Twigs/Buds: Twigs rather stout, smooth, chestnut-brown with white "dots." Leaf scars alternate, half-round, raised. Side buds oval, pointed (to 1/3 inch long), smooth, scales dark brown; end buds sometimes absent.

Flowers/Fruit/Seeds: Flowers unisexual on same tree; appear after leafing in June. Male flowers densely clustered on long (to 8 inches), erect, cream-white, strong-scented stalks; female flowers in small clusters at base of male flower stalks. **Fruits large** (2–3 inches diam.) **spiny burs** with sharp, branched spines, each bur enclosing 1–3 brown, shiny, tasty nuts with a leathery husk. Delicious—eagerly sought by humans and wildlife.

Habitat: Once important in rich upland forests of mixed hardwoods.

Range: Formerly in hill country of southern Indiana. Also planted widely for wood and nuts in early days. Found today largely as stump sprouts; occasionally reaches flowering size.

Comments: Wood soft, lightweight, extremely durable. Once used for furniture, posts, poles, cabin logs. Species essentially wiped out by an introduced fungal pathogen *Cryptonectria parasitica*, the Chestnut blight.

WHITE OAK

Quercus alba L.

Distinguishing Features: Large trees (to 100 feet tall, 4 feet diam.) with broad rounded crown of heavy branches. **Leaves alternate, with rounded, deep lobes. Terminal buds rounded, clustered. Pith star-shaped. Fruit an acorn, oblong, greenish-brown.**

Leaves: Alternate on twig, simple, borne on short (<1 inch), stout, grooved leafstalks. **Blades widest above middle** (to 9 inches long, 1/2 as wide), **divided into 7–9 (rarely 5) shallow to deep rounded lobes, tapering to base; edges smooth.** Bright green, smooth, and shiny above; paler and nearly hairless below. Turn deep red in autumn.

Bark: Thick, light ash gray, with shallow fissures separating long, loose vertical scales or rounded ridges; becoming blocky on old trees.

Twigs/Buds: Twigs slender, greenish, shiny with a white waxy frosting when young, then red-brown later. Pith star-shaped in cross-section. Buds nearly round (<1/4 inch), scaly, red-brown, clustered near twig tip.

See photos at right: Black oak, top; Red oak, middle; White oak, bottom.

Flowers/Fruit/Seeds: Flowers unisexual, borne on same tree as leaves unfold. Male flowers in long (to 3 inches) slender, smooth, yellow, hanging clusters (catkins); female in tiny clusters (of 2–4) at junction of leafstalk and twig. **Fruits are acorns** maturing in one season, borne 1–3 with or without stalk; to 3/4 inch long, greenish to light brown; **cup covering to 1/4 of nut, scales warty.** Seeds mildly bitter, edible; important wildlife food.

Habitat: Wide range of conditions from flatwoods, to moist forests, to wooded slopes, to dry uplands.

Range: Throughout Indiana.

Comments: Wood is heavy, hard, strong, durable, pale brown with an attractive grain. The most important oak species in North America. Used for cabinets, flooring, veneer, office furniture, interior finishing, whiskey barrels, and fuel; formerly in shipbuilding.

POST-OAK

Quercus stellata Wangenh.

Distinguishing Features: A **small to medium tree** (to 60 feet tall, 1–2 feet diam.) with a rounded crown of a few large branches. **Leaves leathery, somewhat cross-shaped with two large, squarish side lobes. Buds clustered at twig tip; pith star-shaped.**

Leaves: Alternate on twig, simple (to 6 inches long, 2/3 as wide), borne on stout, often hairy leafstalk (to 1 inch long). Blades thick, leathery, **squarish upper lobes extend to form a cross; edges of lobes rounded, smooth.** Dark green and hairy above, pale and yellow-hairy below. Turn red to brown in autumn.

Bark: Gray to light brown, divided into deep fissures and scaly ridges.

Twigs/Buds: Twigs stout, brownish, fuzzy early, darker and smooth later. Leaf scars clustered near twig tip; pith star-shaped in cross-section. Buds round (1/8 inch) in cluster at twig tip; hairy, with dark brown scales.

Flowers/Fruit/Seeds: Flowers unisexual on same tree, appear as leaves open. Male flowers in yellow, hanging, slender clusters (3–4 inches long); female flowers tiny, hairy, in clusters at leaf bases. Fruits are small **acorns (to 3/4 inch)** which mature in one year; **top-shaped cup encloses 1/3 of nut, cup red-brown, hairy.** Seed sweet, edible.

Habitat: Usually found on poor soils. Grows in both wet, spring-ponded flatwoods and in dry woods on steep slopes and bluffs.

Range: **Largely in the southwestern quarter of Indiana.**

Comments: Wood is hard, heavy, tough, and strong; light brown. Once used extensively for fence posts (hence common name). Also used for fuel and general construction. Of minor importance due to small size; lumped with White oak when sold.

OVERCUP-OAK

Quercus lyrata Walter.

Distinguishing Features: Medium to large trees (to 80 feet tall, 2–3 feet diam.) with a rounded crown of crooked branches, the lowermost ones often drooping. **Twigs with cluster of terminal buds; pith star-shaped. Spherical acorns have nut almost completely covered by the cup (hence common name). Densely hairy lower leaf surfaces.**

Leaves: Alternate on twig, simple, borne on short (to 1 inch), stout, hairy leafstalks. Blades to 10 inches long, less than half as broad; **edges divided into 5–7 irregular rounded lobes;** tip rounded, tapering to base. Dark green and shiny above, paler and densely hairy below. Turn yellow-brown in autumn.

Bark: Gray-brown, divided into scaly ridges; sometimes in squarish plates; older trees may have patchy areas where bark is sloughed off.

Twigs/Buds: Twigs slender, green and hairy early; then smooth and orange-brown. Leaf scars alternate, half round, clustered near tip. Pith star-shaped in cross-section. Buds clustered at twig tip, nearly round, small (<1/4 inch), smooth, blunt, pale brown.

Flowers/Fruit/Seeds: Flowers unisexual on same tree; appear as leaves unfold. Male flowers in long (4–6 inches), slender, drooping, hairy clusters; female tiny, in small clusters at branch tips. **Fruits are acorns** maturing in one season; the pale brown nut is nearly spherical (to 1 inch); **almost completely enclosed by the deep, round, knobby cup.** Seeds semi-sweet, edible.

Habitat: Wet sites in floodplains, forested swamps, and sloughs. Very tolerant of extended flooding.

Range: Few counties of far southwestern Indiana.

Comments: Wood heavy, hard, strong (may be stronger than White oak), dark brown. Sold as White oak for similar uses, but of inferior quality.

BUR-OAK

Quercus macrocarpa Michx.

Distinguishing Features: Large trees (to 100 feet tall, 4 feet diam.) with a rounded crown of thick branches. **Large leaves with rounded lobes; cut nearly to mid-vein near center of leaf. Acorns very large with hairy fringe at top of cup. Buds clustered at twig tip; pith star-shaped.**

Leaves: Alternate on twig, single, borne on stout, flattened and grooved leafstalks (to 1 inch long). Blades are thick, leathery, to 12 inches long, half as broad, widest above the middle; divided into 5–7 rounded lobes; rounded at tip, tapering to base. Dark green and smooth above, paler and downy below. Turn yellow-brown in autumn.

Bark: Gray-brown, thick, divided into deep furrows and broad ridges on large trees.

Twigs/Buds: **Twigs stout, brown, sometimes with corky ridges.** Leaf scars alternate, clustered near tip, half-round. Pith star-shaped in cross-section. Buds small (1/8 inch), oval to round, pointed, with red-brown hairy scales.

Flowers/Fruit/Seeds: Flowers unisexual on same tree as leaves unfold. Male flowers in long (3–6 inches), slender, yellow, hairy, drooping clusters. Female flowers tiny, reddish, in small clusters near twig tips. **Fruit a large acorn (to 1 3/4 inches long), the deep cup of which covers 1/2 to 3/4 of the nut has loose scales and a bushy, hair-like fringe.** Acorns often have long stalks (to 2 inches). Seed semi-sweet, edible.

Habitat: Does best on fertile bottomland soils and wet-site woods, but also grows on dry limestone ridges.

Range: Throughout most of Indiana on suitable sites. Less common in southern Indiana hill country.

Comments: Medium brown wood is hard, heavy, strong, durable, and tough. Sold and used as White oak. Also called Mossycup oak.

SWAMP WHITE OAK

Quercus bicolor Willd.

Distinguishing Features: Medium to large trees (to 90 feet tall, 3 feet diam.) with broad, rounded, often irregular, crown. **Leaves coarsely round-toothed; dark green and lustrous above, white and softly hairy below** (hence specific name, *bicolor*). **Acorn stalk 1 to 3 inches long.**

Leaves: Alternate on twig, simple, borne on stout leafstalks to 1 inch long. Blade broadly oval, widest above the middle (to 8 inches long, 1/2 as broad); edges coarsely round-toothed; tip rounded, base wedge-shaped. Dark green and shiny above; white and downy below. Leaf veins continue to leaf notches and/or teeth. Turn yellow-brown in autumn.

Bark: Light to dark gray-brown in long loose plates; deeply fissured on old trees; bark on upper large branches scaly, peeling away.

Twigs/Buds: Twigs stout; yellow-brown to red-brown with age, largely smooth. **Pith star-shaped in cross-section.** Leaf scars alternate, raised, half-round. Buds small (<1/4 inch), nearly round, clustered at twig tips; with red-brown scales.

Flowers/Fruit/Seeds: Flowers unisexual on same tree, appear as leaves unfold. Male flowers in long (to 4 inches), slender, drooping clusters; female tiny, in clusters of 2–4 on woolly stalks at branch tips. **Fruit an acorn** maturing in 1 year; nut to 1 inch long **on long stalk** (to 3 inches), often with an aborted second acorn on stalk. Roughened, hairy cup covers 1/3 of nut. Seed semi-sweet, edible; good wildlife food.

Habitat: Bottomland woods and upland wet forests.

Range: Throughout Indiana, but only in selective site conditions.

Comments: Wood heavy, hard, strong, pale brown. Sold as White oak and used similarly, but Swamp white oak lumber is often of inferior quality.

SWAMP CHESTNUT-OAK

Quercus michauxii Nutt.

Distinguishing Features: Medium to large trees (to 110 feet tall, 4 feet diam.) with a broad, rounded crown. **Leaves coarsely round-toothed with densely hairy lower surfaces. Twigs with cluster of terminal buds; pith star-shaped.** Acorn with short stem; thick, bowl-shaped cup with wedge-shaped scales.

Leaves: Alternate on twig, simple, borne on slender, hairy leafstalk (to 1 1/2 inches). Blades oval, broadest above middle (to 9 inches long, 1/2 as wide); edges scalloped with large rounded teeth, decreasing in size toward leaf tip; dark green and shiny above, silvery and densely hairy below. Turn a beautiful crimson red in fall.

Bark: Light gray to yellow-gray; narrowly ridged and scaly on old trees.

Twigs/Buds: Twigs stout, orange-brown to red-brown. Leaf scars alternate; crowded near twig tip. Pith star-shaped in cross-section. Buds oval, pointed, hairy, red-brown, to 1/4 inch, crowded near twig tip.

Flowers/Fruit/Seeds: Flowers unisexual on same tree. Male flowers in long (to 4 inches), hairy, hanging clusters; female tiny, on short, reddish, hairy spikes at leaf bases. **Fruit a red-brown, oval acorn (to 1 1/2 inches long), with a thick bowl-shaped cup which covers 1/3 of nut**—cup has wedge-shaped scales; stalk short or absent. Seed sweet, edible.

Habitat: **Low wet woods and some stream bottom forests. Mostly in upland flatwoods in southeastern Indiana.**

Range: Several county areas in far southeastern and southwestern Indiana.

Comments: Wood hard, heavy, strong, close-grained, pale brown. Sold with White oak for similar uses. Also called Basket oak, for its ease in splitting into strips for baskets; and Cow oak because cattle regularly browse this species. Makes a fine ornamental (with its red fall foliage) and should be used more widely.

ROCK CHESTNUT-OAK

Quercus prinus L.

Distinguishing Features: Medium to large trees (to 80 feet tall, 2–3 feet diam.) with a broad, open, irregular crown. **Leaves thick, leathery, with edges coarsely round-toothed. Bark deeply fissured with age. Buds clustered at twig tips; pith star-shaped. Acorns elongated (to 1 1/2 inches), rich brown.**

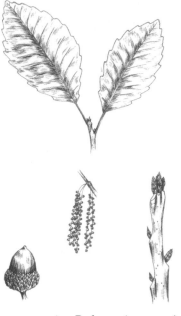

Leaves: Alternate on twig, simple, borne on leafstalks to 1 inch long. **Blades ovate, broadest above the middle, thick,** leathery (to 9 inches long, 1/2 as broad); edges coarsely round-toothed. Yellow-green, shiny and smooth above, paler and finely hairy over all of lower surface. Turn yellow-brown in autumn.

Bark: Dark brown to almost black; thick, hard, with deep fissures between large, rounded, broken ridges on older trees; smooth on young trees.

Twigs/Buds: Twigs stout, orange to red-brown; leaf scars alternate, clustered near twig tip; pith star-shaped in cross-section. **Buds tapering to a point** (to 1/4+ inch), **with shiny brown, hairy scales.**

Flowers/Fruit/Seeds: Flowers unisexual on same tree at time of leafing. Male flowers in hanging, hairy, yellow, slender clusters (to 3 inches); female tiny, on short stalks at leaf bases. Fruits are oval acorns (to 1 1/2 inches long), rich dark brown; cup covering 1/3 to 1/2 of nut, cup scales thickened, warty. Seed sweet, edible.

Habitat: **On well-drained soils of steep ridges and dry, rocky, wooded slopes.**

Range: South-central Indiana hill country and Ohio River border.

Comments: Wood light brown, heavy, strong, tough, and close-grained. Cut and sold as White oak for similar uses. Bark an important tannin source.

CHINKAPIN-OAK

Quercus muehlenbergii Engelm.

Distinguishing Features: Large tree (to 100 feet tall, 4 feet diam.) with rounded crown. **Leaves sharply round-toothed, oval to lance-shaped. Terminal buds clustered; pith star-shaped in cross-section. Acorn egg-shaped, very dark (often nearly black) with thin cup of medium depth.**

Leaves: Alternate on twig, simple, borne on slender leafstalks (to 1 inch). Blades tough, oval (to 7 inches long, 1/2 as wide); **edges coarsely toothed with a tiny gland at tip of each tooth.** Dark green, smooth, and shiny above, pale and finely hairy below. Turn yellow-brown in autumn.

Bark: Light gray with a faint yellowish cast (hence common name); shallowly fissured with scaly ridges.

Twigs/Buds: Twigs slender, yellow-brown to orange-brown. Pith star-shaped in cross-section. Leaf scars alternate, half round, clustered near twig tip. Buds clustered at twig tip; oval (to 1/4 inch), pointed, smooth; scales chestnut-brown.

Flowers/Fruit/Seeds: Flowers unisexual on same tree at time of leafing. Male flowers in hairy, yellow, hanging clusters (to 4 inches long); female tiny on short, hairy spikes at base of leaves. **Fruit an egg-shaped acorn (to 3/4 inch long); nut dark brown to almost black; cup thin-walled, covers 1/3 to 1/2 of nut, scales hairy.** Matures first season. Seed sweet, most edible of all oaks in Indiana.

Habitat: Moist, fertile, upland woods; dry hillsides, especially on limestone soil. Also on gravelly bluffs of major rivers.

Range: Most of Indiana on suitable sites. Uncommon to absent in northwestern counties.

Comments: Wood heavy, hard, strong, durable; heartwood dark brown. Sold as White oak for similar uses. Leaves resemble those of American chestnut. A nice ornamental tree, although rarely used as such. Also called Yellow oak, especially by foresters. Pioneers called it Pigeon oak, presumably for the use of the acorns by passenger pigeons for food.

SHINGLE-OAK

Quercus imbricaria Michx.

Distinguishing Features: Medium tree (to 70 feet tall, 2–3 feet diam.) with a pyramid-shaped to rounded crown with many branches. **Only oak native to Indiana with leaves having no lobes or teeth along the edges. Pith star-shaped. Terminal buds clustered at twig tip.**

Leaves: Alternate on twig, simple, borne on short (to 1/2 inch), stout, hairy leafstalks. Blades to 6 inches long, 2 inches wide, oblong; **edges smooth,** except seedlings sometimes have toothed leaves. Dark green and glossy above, paler and velvety below. Turn rusty brown in autumn.

Bark: Dark brown to black with shallow fissures between hard flat plates when mature; smooth gray when young.

Twigs/Buds: Twigs slender, dark green, and hairy when young; later smooth and red-brown. Leaf scars alternate, half-round, elevated, clustered near tip. Pith star-shaped in cross-section. Buds oval, pointed, small (to 1/8 inch); with brown scales, hairy at margins.

Flowers/Fruit/Seeds: Flowers unisexual on same tree, appear as leaves unfold. Female flowers tiny, without petals, clustered; male in slender, hanging, hairy clusters (2–3 inches long). **Fruits nearly round (to 1/2 inch) acorns,** dark brown; cups thin, shallow bowl-shaped, with tight red-brown, hairy scales, enclose about 1/3 of nut. Acorns mature in second growth season. Kernel very bitter.

Habitat: Variable. Occurs mostly on sandy ridges or as a pioneer species in northern Indiana. Scattered on ridges, or on moist soils along larger streams in southwestern counties.

Range: Throughout much of Indiana on appropriate sites.

Comments: Wood is hard, heavy, coarse-grained, and light reddish-brown. Was split into shingles by the pioneers for their cabins, hence common name. Now sold as Red oak when cut for lumber. Dark, glossy leaves and symmetrical shape make it an attractive ornamental, but not widely used, likely because of unsightly crown galls, which are often present.

BLACK-JACK OAK

Quercus marilandica Muenchh.

Distinguishing Features: **A small scrubby tree** (<40 feet tall, 1 foot diam.), **usually stunted in appearance** with an open, irregular, contorted crown, often with crooked trunk. **Leathery leaves 3-lobed, much broader toward tip. Pith star-shaped. Bark black, broken into rough blocks on mature trees.** Acorns with loose, red-brown scales.

Leaves: Alternate on twig, simple, borne on short (<1 inch) leafstalks. **Blades thick and leathery with 3 lobes, weakly bristle-tipped; resemble footprint of web-footed waterbird.** Dark green, shiny, and smooth above; yellow-woolly below. Turn yellow to brown in autumn.

Bark: Black, thick on trunk, broken into hard rough square or rectangular blocks when mature.

Twigs/Buds: Twigs stout, red-brown to brown, usually hairy. Leaf scars alternate, half-round, elevated, clustered near tip. Pith star-shaped in cross-section. Buds clustered at twig tip, angular, oval (to 1/3 inch), with rusty-brown hairs on scales.

Flowers/Fruit/Seeds:
Flowers unisexual on same tree, appear as leaves unfold. Female flowers red and green, clustered on short spikes near twig ends; male in hanging, hairy clusters. Fruits are **acorns** without stalks, mature in second season; **nearly round (to 1/2 inch), cup deep, bowl-like, with loose red-brown scales, encloses half of nut.** Kernel bitter.

Habitat: Poor soil, especially on dry, exposed rocky cliffs; in barrens and glades.

Range: Southwestern quarter of Indiana.

Comments: The dark brown wood is heavy, hard, strong and coarse-grained, but it does not make good lumber. Excellent for firewood and charcoal.

170

CHERRYBARK-OAK

Quercus pagoda Raf.

Distinguishing Features: Trees large (to 110 feet tall, 4 feet diam.) with straight, clear trunks and broad rounded crowns. **Leaf lobes not recurved and nearly at right angles to the mid-vein. Bark closely resembles that of wild black cherry** (hence common name).

Leaves: Alternate on twigs, simple, borne on stout, hairy leafstalks (to 2 inches long). Blades divided into **5–11 narrow, long-pointed bristle-tipped lobes; not recurved and nearly at right angles to midline** of leaf (leaves to 9 inches long, 2/3 as wide). Dark green and lustrous above, paler and downy below. Leaves highly variable, even from same tree.

Bark: Dark gray-brown to nearly black; broken into shallow fissures and cross cracks, resulting in a **scaly to blocky** appearance that **resembles the bark of wild cherry.**

Twigs/Buds: Twigs medium stout, red-brown; downy early, then smooth. Leaf scars alternate, half-round, elevated, clustered near twig tip. Pith star-shaped in cross-section. Buds oval, pointed, angular, to 1/4 inch long, with hairy, chestnut-colored scales.

Flowers/Fruit/Seeds: Flowers unisexual on same tree, usually appear before leaves unfold. Female flowers few, on stout, hairy stalks near base of leafstalks; male in drooping, hairy clusters (to 5 inches long). **Fruits are nearly round (about 1/2 inch) acorns** which mature in 2 years; **cups enclose about 1/3 of nut, scales hairy.** Kernel bitter.

Habitat: **Usually in bottomland forests or swamps and on flats along streams;** occasionally on drier sites.

Range: Southwestern Indiana.

Comments: Wood is light reddish-brown, heavy, hard, strong, and coarse-grained; of very good quality. Used for interior finishing, furniture, cabinets, and flooring.

SOUTHERN RED OAK

Quercus falcata Michx.

Distinguishing Features: A medium to large tree (to 80 feet tall, 3 feet diam.) with a rounded to spreading crown of stout branches. **Leaves somewhat bell-shaped *or* with narrow, strap-like, sickle-shaped lobes; lower leaf surface with a mat of fine rust-colored hairs. Acorns to 1/2 inch long, cup shallow.**

Leaves: Alternate on twig, simple, borne on long (to 2 1/2 inches) slender, hairy leafstalks. **Blades with 3 to 5 bristle-tipped lobes,** broadest above middle (to 8 inches long, 1/2 as wide); highly variable between two types: one with slender, pointed, strap-like lobes ("turkey-track" shaped); the other somewhat bell-shaped (both have **sickle-shaped lobes,** hence scientific name). Shiny dark green above, paler with rusty hair below.

Bark: Light gray and smooth when young; dark brown to black, rough and fissured with broad, hard ridges when mature.

Twigs/Buds: Twigs are red-brown to dark brown; rusty-hairy early, then smooth. Leaf scars alternate, half-round, clustered near tip. **Pith star-shaped in cross-section.** Buds oval, pointed; chestnut-brown, hairy scales; to 1/4 inch long.

Flowers/Fruit/Seeds: Flowers unisexual on same tree, appear as leaves unfold. Female flowers reddish, singly or paired on short, hairy stalks; male in drooping, hairy clusters (3–5 inches). Fruits are round, **orange-brown acorns to 1/2 inch long; borne in pairs; scaled cups are saucer-shaped, cover up to 1/3 of nut.** Kernel bitter.

Habitat: Usually on dry, poor upland ridges, but also in flatwoods and bottomland forests.

Range: Far southern and southwestern (especially) Indiana.

Comments: Reddish-brown wood is hard, strong, and coarse-grained, but is inferior in quality to Northern red oak. Used for rough construction, fuel, pallets, and crating. Makes a nice ornamental, but not commonly used as such. Also called Spanish oak.

BLACK OAK

Quercus velutina Lam.

Distinguishing Features: Trees of medium to large size (to 90 feet tall, 3 1/2 feet diam.) with a broad, rounded crown. **Terminal buds clustered at twig tip are large, angular, and densely gray-hairy. Pith is star-shaped. Acorn cup, which encloses 1/2 of nut, has loose, fringed scales.**

Leaves: Alternate on twig, simple, borne on stout, long (to 4 inches) leafstalks. **Blades leathery, deeply divided into 5–9 bristle-tipped lobes;** to 8 inches long, 2/3 as broad, widest above the middle, sharp-pointed at tip, unevenly squared at base. Dark green, glossy, and smooth above; yellow-green with rust-colored hair on veins below. Turn yellow to russet-brown in autumn.

Bark: Smooth and brown-black on young trees; with age becomes hard, black, deeply furrowed and scaly to rough-blocky at the base. Compare photo of Black oak on p. 177 to Northern red oak on p. 179. **Inner bark a bright yellow-orange.**

Twigs/Buds: Twigs moderately stout; angled, smooth, dark reddish-brown when mature. Leaf scars alternate, half-round. Pith star-shaped in cross-section. **Buds large (to 1/3 inch), clustered at branch end, oval, pointed, strongly angled; covered with gray wool.**

Flowers/Fruit/Seeds: Flowers unisexual on the same tree; appear as leaves unfold. Female flowers tiny, red, clustered near base of leafstalks; male in hanging, hairy, long (4–6 inches) clusters (catkins). Fruit an **acorn** maturing in 2 years; **3/4 inch long, nearly round; cup bowl-shaped, encloses 1/2 of nut, scales loose, ragged.** Kernel very bitter.

Habitat: Dry upland forests, upper slopes, and dry rocky or sandy ridges.

Range: Throughout Indiana.

Comments: The reddish-brown wood is hard, heavy, strong, and close-grained. Used for railroad ties, flooring, rough construction. Now sold as Red oak for similar uses. Tannin once obtained from inner bark. Not used ornamentally because of its poor form, appearance.

176

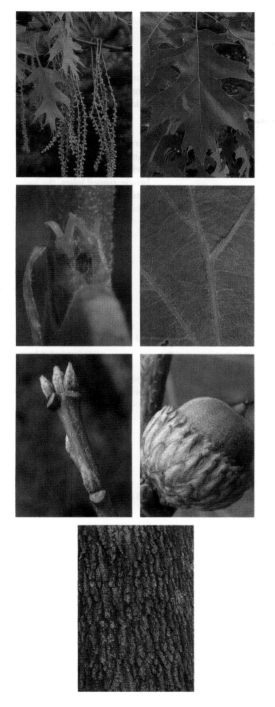

NORTHERN RED OAK

Quercus rubra L.

Distinguishing Features: Large tree (to 100 feet tall, 3 1/2 feet diam.), trunk straight; crown spreading, rounded. **Bark ridges are smooth, shiny continuous "tracks" running from large branches to base of trunk. Acorns very large with a shallow, saucer-like cup. Inner bark pink to tan.**

Leaves: Alternate on twig, simple, borne on slender, smooth leafstalks (to 2 inches long). **Blades shallowly divided into 7–11 upward-pointing, bristle-tipped lobes, the indentations extending only half way to the central vein;** to 8 inches long, 2/3 as wide, broadest above the middle. Dark green, smooth, and shiny above; paler and largely smooth below. Turn brilliant red in autumn.

Bark: Dark brown to black. Smooth when young; shallowly furrowed with broad smooth streaks on vigorous trees; deeply furrowed on old trunks.

Twigs/Buds: Twigs slender, smooth, reddish-brown. Leaf scars alternate, half-round, clustered near twig end. **Pith star-shaped in cross-section. Buds pointed (to 1/4 inch long), the smooth, shiny red-brown scales** sometimes hairy.

Flowers/Fruit/Seeds: Flowers unisexual on same tree; appear as leaves unfold. Female flowers tiny, clustered near base of leafstalks; male in slender, hanging, hairy clusters (catkins) (4–5 inches long). **Fruit a large (to 1 1/4 inches long), robust acorn** which matures in 2 years; **cup saucer-shaped, covers <1/4 of nut, scales tight, reddish brown.** Kernel somewhat bitter.

Habitat: Rich, upland woods, well-drained north- or east-facing slopes, and occasionally along river banks. Not usually found on wet or poorly drained soils.

Range: Throughout Indiana.

Comments: The light reddish-brown wood is hard, heavy, strong, and coarse-grained. Used for interior finishing, flooring, furniture, cabinets, fuel, and construction lumber. The most important species of the red oak group. Excellent growth form and lovely fall color make it a highly desirable ornamental.

PIN-OAK

Quercus palustris Muenchh.

Distinguishing Features: Medium to large trees (to 90 feet tall, 2–3 feet diam.); a broad pyramid-shaped crown with **drooping lower branches that prune poorly. Leaves small, with bristle-tipped lobes at right angles to leaf axis. Cluster of buds at twig ends; pith star-shaped. Acorns small (to 1/2 inch), nearly round, cup shallow.**

Leaves: Alternate on twig, simple, borne on slender, smooth leafstalks (to 2 inches long). Blades small (to 6 inches long, 2/3 as broad), **5–7 bristle-tipped lobes nearly at right angles to leaf axis, lobes diverge from one another.** Dark green, shiny, and smooth above, paler below with hair tufts in vein angles. Turn russet-brown in autumn.

Bark: Thin, smooth, and dark brown when young; slightly furrowed and dark gray-black on older trees.

Twigs/Buds: Twigs slender, smooth, red-brown to dark gray. Leaf scars alternate, crowded near tip. Pith star-shaped in cross-section. Buds clustered at twig end, tapered, pointed, reddish-brown, smooth (to 1/8 inch).

Flowers/Fruit/Seeds: Flowers unisexual on same tree, appear as leaves unfold. Female flowers tiny in clusters of 2–4 near ends of leafstalks; male flowers are in hanging, hairy, clusters (to 3 inches long). Fruits are **nearly round, small acorns (to 1/2 inch),** which mature in 2 seasons; **cup shallowly saucer-shaped, scales thin; nut striped.** Kernel bitter.

Habitat: Moist soil of floodplains, flatwoods, along edges of swamps, sloughs, or ponds.

Range: Throughout most of Indiana.

Comments: Its light brown wood is coarse-grained, hard, and heavy, but not as strong as other species in red oak group. Inferior for lumber since non-pruning branch bases (pins) continue into heart of tree, hence common name of tree. Used for fuel and rough construction. An excellent and widely used ornamental.

SHUMARD OAK

Quercus shumardii Buckley.

Distinguishing Features: Large tree (to 110 feet tall, 4+ feet diam.) with straight trunk and broad, open, rounded crown. **Leaves deeply lobed, shiny, with broadly rounded "C-shaped" indentations. Acorns similar to those of Northern red oak, but slightly smaller.** Buds smooth, straw-colored.

Leaves: Alternate on twig, simple, borne on slender, smooth leafstalks (to 2 1/2 inches long). Blades large (to 8 inches long, 2/3 as wide) with **7–9 bristle-tipped lobes; divided 2/3 of the way to the central vein, the indentations rounded, "C-shaped"**; nearly square at base, tip sharply pointed. Dark green, shiny, and smooth above, paler below, with white tufts of hair at vein angles. Turn red to russet in autumn.

Bark: Smooth and dark gray to black on young trees; shallowly furrowed and ridged on vigorous trees; deeply fissured, hard, and rough on old trunks.

Twigs/Buds: Twigs moderately stout, smooth, tan to reddish. **Pith star-shaped in cross-section.** Leaf scars clustered near tip, half-round, elevated. Buds clustered at twig tip, oval, pointed, smooth, straw-colored, to 1/4 inch long.

Flowers/Fruit/Seeds: Flowers unisexual on same tree; appear as leaves unfold. Female flowers tiny, in clusters near end of leafstalks; male in slender, hairless, hanging clusters (5–7 inches long). Fruits are pale brown **acorns** maturing in 2 seasons; **to 1 inch long, robust; cup deeply saucer-shaped, enclosing 1/4–1/3 of nut, scales thin, overlapping, free from cup.** Kernel bitter.

Habitat: Usually in bottomland woods and upland depressional forests. Sometimes in dry upland sites over limestone in southeastern Indiana.

Range: Southern three quarters of Indiana definitely; likely also occurs in many of the northern tiers of counties.

Comments: Wood is light reddish-brown, hard, heavy, close-grained, and valuable. Equal or superior to Northern red oak in quality and used similarly. A handsome ornamental that should gain wider usage.

NORTHERN PIN-OAK

Quercus ellipsoidalis E. J. Hill.

Distinguishing Features: Trees of small to medium size (to 70 feet tall, 1–2 feet diam.) with a rounded crown and **drooping non-pruning lower branches that nearly reach the ground. Leaf lobes bristle-tipped with openings cut almost to mid-vein. Acorns football-shaped with bowl-shaped cup enclosing 1/3 to 1/2 of the nut.**

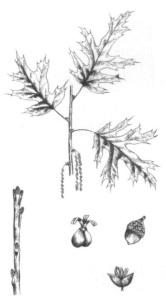

Leaves: Alternate on twig, simple, borne on long (to 2 1/2 inches), slender, smooth leafstalks. **Blades (to 6 inches long, almost as wide) divided into 2–4 pairs of bristle-tipped lobes, the narrow C-shaped openings extending 3/4 of way to mid-vein;** base of leaf nearly straight across. Deep green, shiny, and smooth above; paler below with hair tufts in vein angles. Turn russet-brown in autumn.

Bark: Gray-black; smooth on young trees; developing shallow fissures and hard, narrow ridges with age.

Twigs/Buds: Twigs slender, smooth, red-brown to gray-brown; leaf scars alternate, clustered at twig tip; **pith star-shaped in cross-section.** Buds clustered at twig end, oval (to 1/4 inch), tapering; shiny red-brown scales.

Flowers/Fruit/Seeds: Flowers unisexual on same tree, appear as leaves unfold. Female flowers tiny, in clusters, at leafstalk bases; male flowers in slender, hanging, hairy clusters (to 2 inches long). Fruits are acorns maturing in two seasons; **nut oval, striped (to 3/4 inch), cup is bowl-shaped, covers 1/3 to 1/2 of nut, cup scales thin.** Kernel very bitter.

Habitat: Sandy ridges, dry upland woods and adjacent lowlands.

Range: **Northern quarter of Indiana,** primarily on glacial outwash areas.

Comments: The pale brown wood is hard, heavy, and strong, but of inferior commercial quality. Sold with Red oak for similar uses, whenever harvested. Could be used ornamentally. Also called Hill's oak. (See also "Species Excluded," p. 337.)

SCARLET OAK

Quercus coccinea Muenchh.

Distinguishing Features: Tree of medium size (to 70 feet tall, 1–2 feet diam.) with a narrow, open crown; trunk tall, straight. **Leaves with bristle-tipped lobes divided almost to leaf center by deep C-shaped notches (sinuses). Buds clustered at twig tips; pith star-shaped. Acorns (to 3/4 inch) usually have concentric rings around tip of nut.**

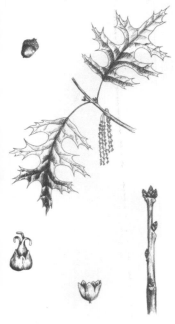

Leaves: Alternate on twig, simple, borne on slender leafstalks (1 1/2–2 1/2 inches long). Blades (to 6 inches long, 2/3 as wide) are divided nearly to center into 5–7 bristle-tipped lobes; the openings between lobes deeply C-shaped. Bright green, shiny, and smooth above; paler and smooth below, with hair tufts at vein angles. Turn brilliant scarlet in autumn (hence common name).

Bark: Nearly black, divided into rough, irregular, scaly, hard ridges.

Twigs/Buds: Twigs slender, smooth, brown; leaf scars alternate, crowded at twig tips; pith star-shaped in cross-section. **Buds** oval (to 1/4 inch), tapering, red-brown scales **covered by fine, gray hair.**

Flowers/Fruit/Seeds: Flowers unisexual on same tree; appear as leaves unfold. Female flowers tiny, clustered on short spikes near leaf bases; male flowers in slender, smooth, hanging clusters (3–4 inches long). Fruits are acorns maturing in two years—to 1 inch long, cup thin, **bowl-shaped, scales tight; nut usually has concentric rings around top.** Kernel bitter.

Habitat: **Dry upland woods and steeper slopes in southern Indiana; on sandy ridges in northern part of state.**

Range: South-central hill country of Indiana and scattered in counties of northern quarter of state.

Comments: Wood hard, heavy, strong, pale brown. Uses similar to those of Red oak and sold with it. A valuable and much sought ornamental for its pleasing growth form and striking fall coloration. May retain dead limbs on trunk.

HOP-HORNBEAM

Ostrya virginiana (Mill.) K. Koch.

Distinguishing Features: Small tree (to 35 feet tall, <1 foot diam.) with irregular, rounded, open crown; usually in forest understory. **Bark brown, shreddy to scaly. "Hop-like" fruits resemble Japanese lanterns. Margins of thin leaves have many fine teeth.**

Leaves: Alternate on twig, simple, borne on very short (to 1/4 inch), slender, hairy stems. Blades thin, oval (to 5 inches long, 1/2 as wide), long-pointed tip; **edges finely double-toothed;** dark yellow-green and smooth above, paler below with hair tufts at vein angles. Turn yellow-brown in autumn.

Bark: Rich brown, smooth on young branches and small trees. Brown and breaking into short, flat strips at maturity—giving a shreddy or scaly appearance to large trees.

Twigs/Buds: **Twigs slender, rich brown, hard to break.** Leaf scars alternate, elevated, crescent-shaped. End buds absent. Side buds small (<1/4 inch), slender, pointed, with overlapping red-brown scales.

Flowers/Fruit/Seeds: Unisexual flowers borne separately on same tree. Male flowers in long (to 2 inches), slender, drooping clusters of three; female in small (1/3 inch) reddish clusters at branch tips. **Fruits, which resemble elongated clusters of hops,** have small (<1/3 inch) flattened seeds enclosed in papery, inflated husks.

Habitat: Upland woods, hillsides, and rocky slopes, usually under a taller forest canopy.

Range: Throughout Indiana.

Comments: Wood heavy, very strong, tough, close-grained, light red-brown. Used for tool handles, levers, and mallet heads by pioneers because it is so strong. Also called Ironwood, an appropriate name.

HORNBEAM; BLUE BEECH; MUSCLEWOOD

Carpinus caroliniana Walter.

Distinguishing Features: Small tree (to 30 feet tall, 1 foot diam.) with irregular, rounded crown; usually in forest understory. **Bark smooth, blue-gray; trunk ridged, appearing "muscular." Fruits are nutlets with attached leafy bracts. Leaves alternate, thin, doubly toothed.**

Leaves: Alternate on twig, simple, borne on slender, hairy stem (<1/2 inch). Blades thin, oval (to 4 inches long, 1/2 as wide), pointed tip; **finely double-toothed;** blue-green and smooth above, pale yellow-green below with hair tufts at vein angles. Turn yellow in autumn.

Bark: Thin, smooth, dark bluish-gray, ridged, appearing muscular (hence common name of Musclewood). Trunks often mottled with lighter or darker patches.

Twigs/Buds: **New twigs slender, hairy, light green to gray,** difficult to break; becoming lustrous red-brown in winter. Leaf scars alternate, elevated, crescent-shaped. End bud absent; side buds small (<1/4 inch), pointed, often 4-angled, with red-brown, downy scales.

Flowers/Fruit/Seeds: Unisexual flowers borne separately on same tree. Male flowers in drooping clusters (to 1 1/2 inches long); female in hairy clusters (to 1/2 inch) at branch tips. **Fruits are clusters of small (to 1/3 inch) nutlets, each with an attached 3-lobed, leafy bract.**

Habitat: Moist woods and stream bottoms, usually in the shade of taller trees.

Range: Throughout Indiana.

Comments: Wood heavy, hard, and very strong, hence another common name, Ironwood. Small tree size limits use to tool handles, mallets, levers, etc. Pioneers used it for "treenware" (bowls, dishes, mugs, etc.) and ox-yokes.

Often has a distinctive lichen growing on trunk. Also called Ironwood and Water beech.

YELLOW BIRCH

Betula alleghaniensis Britton.

Distinguishing Features: Small to medium tree (to 50 feet tall, 1+ foot diam. in Indiana) with rounded, twiggy crown. **Bark silvery to yellowish (hence common name). Leaves alternate, simple. Long, drooping male flowers; fruits cone-like.**

Leaves: Alternate on twig, simple, borne singly on short (to 1 inch) stout, yellow, hairy leafstalks. Blades oval, to 5 inches long, 1/2 as wide, finely double-toothed; dark green and smooth above, paler and somewhat hairy below. Turn bright yellow in autumn.

Bark: **Thin, silvery to yellowish, even bronze, horizontally peeling and curling when young. Becomes very rough, scaly, yellow-gray with age.**

Twigs/Buds: Twigs slender, green-brown, smooth, with many long whitish pores. Have faint odor of wintergreen when bruised. End bud absent. Side buds small (<1/4 inch), sharply pointed, with hairy, rusty-brown scales.

Flowers/Fruit/Seeds: Unisexual flowers appear with leaves in April–May; male and female on same tree. Male flowers in slender, drooping, long (to 3 inches) clusters; female in upright dense (to 3/4 inch) clusters near branch tips. **Broad cone-like fruits (to 1 1/2 inches) are composed of many 3-lobed scales, each covering a tiny, winged seed.** Seed is wind-scattered.

Habitat: Local in Indiana in bogs, lake margins, and along a few streams in northern counties. In a few moist ravines in southern Indiana.

Range: Northern tiers of counties and Crawford County.

Comments: Wood is heavy, hard, strong, and pale brown. Used for paneling, furniture, cabinets, and flooring, but no commercial stands occur in Indiana. Characteristic tree of northern lakes and bogs.

RIVER BIRCH

Betula nigra L.

Distinguishing Features: A medium-sized tree (to 75 feet tall, 2+ feet diam.) with broad, irregular, twiggy crown. **Its shaggy, peeling, reddish-brown bark is distinctive from any other Indiana tree. Leaves alternate, simple, squarish, and double-toothed.** Often has multiple trunks in clumps.

Leaves: Alternate on twig, simple, borne singly on short (<1/2 inch), slender, woolly stems. Blades oval to rhombic (squarish), to 3 inches long, edges coarsely doubly toothed; dark green and shiny above, paler and densely hairy below; rough-textured. Turn yellow-brown in autumn.

Bark: **Thin and papery in pinkish-tan to reddish-brown curls on young trees and branches; thicker, nearly black and platy when mature.** Very distinctive, even at a distance.

Twigs/Buds: Twigs slender, shiny red-brown with alternate, narrow leaf scars. Buds small (to 1/4 inch), red-brown, sharp-pointed, with slightly hairy scales.

Flowers/Fruit/Seeds: Flowers unisexual, male and female on same tree; open April–May. Male flowers on long (1 1/2–3 inches) drooping clusters (catkins); female in short (<1/2 inch), woolly clusters. **Cone-like cylindrical fruits (to 1 inch long and 1/2 inch thick), each contain many tiny 3-lobed, hairy, winged, nut-like seeds.** Seeds wind-scattered in second spring.

Habitat: Along rivers, streams, and in wet-site woods.

Range: Primarily in southwestern third and northwestern quarter of state; occasional in southeastern Indiana. Largely absent elsewhere unless planted.

Comments: Lightweight but strong, wood is a warm pale brown; pretty when finished—sometimes used for furniture, cabinets. Becoming more common as an ornamental. Also called Red birch.

WHITE BIRCH; PAPER BIRCH

Betula papyrifera Marshall.

Distinguishing Features: Medium tree (to 60 feet tall, 1–2 feet diam.) with an irregular small crown of slender, twiggy branches. **White bark peels off into thin, papery layers. Often has multiple trunks in clumps. Long, drooping male flowers, fruits cone-like.**

Leaves: Alternate on twig, simple, borne singly on medium length (to 1 inch), yellow, stout, smooth to hairy, stems. Blades to 3 inches long, 1/2 as wide, triangular to egg-shaped, pointed at tip, rounded at base, coarsely toothed edges, thick-textured. Dark green and smooth above; paler, downy, and with black dots below. Turn yellow in autumn.

Bark: Thin, orange-brown on very young trees; then **white or creamy, and splitting into papery, peeling layers;** nearly black and furrowed at the base of old trees.

Twigs/Buds: Twigs slender, zigzag, dark reddish-brown, somewhat hairy. Leaf scars alternate, crescent-shaped. Side buds small (to 1/4 inch), slender, smooth, pointed, dark brown to nearly black scales.

Flowers/Fruit/Seeds: Unisexual flowers develop separately on the same tree in autumn, remain on tree in winter, and open in spring. Male flowers in long (to 4 inches), slender, brown spikes; female flowers in erect clusters (to 1 1/2 inches) near branch tips. **Fruits cone-like, cylindrical, drooping with many tiny 3-lobed, winged seeds.** Seeds wind-scattered.

Habitat: Fertile forested slopes, stream banks, and wetland borders.

Range: In Indiana it occurs in only a few counties in the northwestern corner of the state.

Comments: Widely scattered and of minor importance in Indiana plant communities. Wood lightweight but strong, close-grained, light reddish-brown. Used for paper pulp, fuel, toothpicks. Bark once used for canoes and tinder. Frequently planted as an ornamental. Also called Canoe birch.

196

BASSWOOD

Tilia americana L.

Distinguishing Features: Medium to large trees (to 80 feet tall, 3 feet diam.) with broad crown; **commonly has root sprouts around trunks of mature trees. Bark deeply furrowed. Leaves heart-shaped, toothed. Buds red in winter.**

Leaves: Alternate, simple, borne singly on long (1 1/2– 2 1/2 inches) stems. Blades oval to heart-shaped (3–6 inches long, 2/3 as wide), **toothed edges, base not symmetrical;** green and smooth above, paler and somewhat hairy below. Turn yellow in autumn.

Bark: Smooth on young trees; dark gray and deeply fissured when trees mature; resembles white ash or black walnut.

Twigs/Buds: Twigs slender, zigzag, greenish to dark gray, leaf scars half round. **Buds rounded, pointed tip, to 1/4 inch long; shiny bright red in winter with two visible scales.**

Flowers/Fruit/Seeds: Bisexual flowers yellow-white, fragrant, in loose clusters on long stalk that has a leafy wing; appear June/July. **Fruits in clusters, hard, nearly round (to 1/3 inch), nutlike, woolly; with leafy wing remaining;** ripe in September/October; wind-scattered.

Habitat: Rich, moist woods, especially in ravines.

Range: Throughout Indiana.

Comments: All native basswoods of Indiana are included herein as *Tilia americana*. Formerly, White basswood, *T. heterophylla*, was considered to be a separate species. (See also "Species Excluded," p. 337.) Wood nearly white to pale brown, soft, smooth grained, and easily worked. Excellent for carving; also used for beekeeping materials, cabinetry, window sashes, and venetian blinds. Honeybees actively seek flowers and produce basswood honey, a delicacy. Also called American linden, or Linn.

BIG-TOOTHED ASPEN; LARGE-TOOTHED ASPEN

Populus grandidentata Michx.

Distinguishing Features: Medium tree (to 60 feet tall, 1 1/2 feet diam.) with a small rounded crown. **Leaves nearly circular with large, coarse teeth (hence common and specific names), on long flattened stalks;** leaves rustle in breezes.

Leaves: Alternate on twigs, simple, borne on long (to 3 inches), slender, flat leafstalks; **flutter in breezes.** Blades smooth, thick, nearly round (to 4 inches long), edges very coarsely single-toothed; dark green above, paler below. Turn yellow-gray in autumn.

Bark: Light gray to greenish and smooth when young. Thick, dark brown to black, furrowed, and scaly when mature.

Twigs/Buds: **Twigs stout, gray-green with numerous orange "dots,"** hairy at first, then smooth. Leaf scars alternate, raised, 3-lobed. End buds small (<1/4 inch), pointed light brown, dusty-hairy; side buds smaller, downy, sticky.

Flowers/Fruit/Seeds: Flowers unisexual on separate trees as leaves unfold; occur in long (to 3 inches), drooping, downy clusters. **Fruits in long clusters of downy capsules which split to release numerous tiny, hairy seeds;** wind-scattered in May.

Habitat: On wooded slopes, sandy and gravelly ridges, or woods edges.

Range: Widely scattered throughout most of Indiana; apparently rare on central tillplain; occasional elsewhere.

Comments: Wood lightweight, soft, pale brown. Used primarily for pulpwood. Grows rapidly; short-lived.

QUAKING ASPEN

Populus tremuloides Michx.

Distinguishing Features: Medium tree (to 50 feet tall, 1 1/2 feet diam.) with a small rounded crown. **Leaves nearly round, finely toothed on long flat stalks, which allow the leaves to flutter (quake) in gentle breezes (hence common and specific names).**

Leaves: Alternate on twigs, simple, borne on long (to 3 inches), slender, flat leafstalks. Blades nearly round (to 3 inches) or broadly oval; edges with many small, rounded teeth; upper surface dark green, shiny, and smooth, paler and dull below. Turn a bright yellow in autumn.

Bark: **Creamy-white to pale yellow-green on young trees and branches;** becomes dark gray to almost black in horizontal bands, or at base of trunk, at maturity.

Twigs/Buds: Twigs slender, smooth, yellow-green to almost white. Leaf scars alternate, crescent-shaped. End buds lance-shaped (<1/4 inch), pointed, oval, sticky, with rich brown scales; side buds smaller, downy.

Flowers/Fruit/Seeds: Flowers unisexual on separate trees as leaves unfold; occur in very long (male to 4 inches; female to 5 inches) drooping, crowded clusters. **Fruits in long clusters of light green, downy capsules, splitting in halves to release numerous tiny seeds with cottony hairs attached.** Seeds wind-scattered in spring.

Habitat: Moist sandy or gravelly slopes; woods or wetland edges; **root sprouts into extensive aspen thickets.**

Range: Scattered in northern half of Indiana.

Comments: Wood lightweight, soft, pale brown. Used mainly for pulpwood. An important successional tree following deforestation or fire in habitats favorable to it. Also called Trembling aspen.

SWAMP-COTTONWOOD

Populus heterophylla L.

Distinguishing Features: Large tree (to 100 feet tall, 2 1/2 feet diam.) with an open irregular crown of a few large, upright branches. **Large, thick, oval leaves are woolly white below. Buds large, oval, sticky, dark brown.**

Leaves: Alternate on twig, simple, borne on **very long (to 4 inches), smooth, not flattened leafstalks.** Blades very large (to 8 inches long, 6 inches broad) with rounded teeth on edges; dark green above, paler below and densely white-woolly when young (hence specific name of tree).

Bark: Lightly furrowed, brown to red-brown when young; breaks into long, narrow, loose plates with age.

Twigs/Buds: **Twigs rather slender, hairy when young; smooth, brown, and shiny with orange center when older.** Buds large (to 1/2 inch long), oval, pointed, sticky, with red-brown scales.

Flowers/Fruit/Seeds: Flowers unisexual on separate trees before leaves unfold; occur in very long (male to 4 inches; female to 6 inches) crowded clusters. **Fruits are clusters of reddish-brown capsules (to 1/2 inch long); contain numerous, tiny, red-brown seeds with cottony hairs;** wind- or water-scattered.

Habitat: Swamps, low wet woods on waterlogged soils.

Range: Scattered in southwest third and northeast third of Indiana; apparently absent elsewhere.

Comments: Wood lightweight, soft, pale brown. Used for pulpwood, boxes and crates, interior finishing materials. Handsome tall, straight tree of deep swamp habitats. Cottony seeds sometimes cover the ground "snow-like" in dense stands during heavy seed years.

COTTONWOOD

Populus deltoides Marshall.

Distinguishing Features: Large tree (to 110 feet tall, 6 feet diam.) with a spreading crown of large branches, some drooping. **Leaves are triangular** (hence specific name), **coarsely toothed, and have flattened stems. Buds large, pointed, sticky.**

Leaves: Alternate on twig, simple, borne on long (to 4 inches), slender, flattened, smooth, yellow leafstalks. Blades thick, broad, triangular (to 6 inches long, almost as broad), edges coarsely toothed; dark glossy green above, paler below. **Noisily flutter in breezes.** Turn yellow-gray or pale brown in autumn.

Bark: Smooth and yellow-green when young. Thick, dark gray and deeply furrowed when mature.

Twigs/Buds: Twigs stout, shiny, sometimes with corky ridges, often with numerous pale "dots"; leaf scars alternate, triangular. End buds large (to 3/4 inch), shiny, sticky, long-pointed, with chestnut brown scales; side buds similar but much smaller.

Flowers/Fruit/Seeds: Flowers unisexual on separate trees before leaves unfold. Flowers crowded in long (to 3 inches), thick, drooping clusters (catkins); male clusters red, female yellow-green. **Fruits in long clusters of capsules, which contain numerous seeds with cottony hairs attached (hence common name of tree).** Seeds wind- or water-scattered in spring, sometimes covering the ground with cottony fibers.

Habitat: Bottomland woods along streams; occasionally in swamp forests.

Range: Throughout Indiana.

Comments: Wood lightweight, soft; sapwood large and white, heartwood small and brown. Used for pulpwood, basket veneer, matchsticks, crates and boxes, some furniture. Very fast-growing tree.

BLACK WILLOW

Salix nigra Marshall.

Distinguishing Features: Medium tree (to 60 feet tall, 2+ feet diam.) with a spreading, irregular, twiggy crown. Small branches usually droop. Alternate, narrow lance-shaped leaves. Very slender, brittle twigs; buds with a single scale.

Leaves: Alternate on twig, simple, borne on very short (<1/4 inch) smooth stems, **often with paired leafy stipules ("wings") at the base.** Curved blades long (to 6 inches), narrow (to 3/4 inch), lance-shaped; finely toothed on edges; bright green and shiny above, very pale green below. Turn yellow in autumn.

Bark: Very dark gray-brown to black in long, loose, vertical scales or thin plates. Deeply furrowed and shaggy on old trees.

Twigs/Buds: **Twigs very slender, "whip-like," olive to yellow-green, smooth, brittle.** Leaf scars alternate, U-shaped. Buds small (1/8 inch), cone-shaped, smooth, pointed; with a single scale covering the entire bud.

Flowers/Fruit/Seeds: Flowers unisexual on separate trees in early spring. Male flowers yellow in hairy, crowded, slender, drooping clusters (to 3 inches long); female flowers green, urn-shaped, clusters shorter. **Fruits are very small (to 1/8 inch), egg-shaped, red-brown capsules in clusters; each containing many tiny silky-haired seeds.** Seeds are wind-scattered.

Habitat: Wet sites along stream banks, in low wet woods, and around pond or lake borders.

Range: Throughout Indiana.

Comments: Wood very lightweight, soft, flexible, but not strong. Used for barrels, boxes, crates, charcoal, pulpwood, some furniture, and artificial limbs (in early days). Cuttings used in erosion control of streambanks.

SOURWOOD

Oxydendrum arboreum (L.) DC.

Distinguishing Features: Small to medium-sized tree (to 50 feet tall, 1 foot diam.) with an open crown of crooked branches. **Gray bark tinged with red in the fissures. Long, oval, strap-like, shiny leaves. Abundant, showy white flower-sprays in summer. Growth form and bark resemble sassafras.**

Leaves: Alternate, simple, borne singly on very short (to 1/2 inch) stems. Blades are long ovals (4–6 inches, 1/3 as wide); long tapering tip, narrow at base; edges finely toothed; smooth, shiny green above, paler below; sour to taste. Turn a lovely scarlet in autumn.

Bark: Thick, furrowed with broad, scaly ridges; gray tinged with red in fissures.

Twigs/Buds: Twigs slender, drooping, greenish to red-brown, without hairs. End bud absent; **side buds small, scaly, red-brown, partly embedded in bark.**

Flowers/Fruit/Seeds: Bisexual flowers appear in June after leafing. Many, small, urn-shaped **white flowers** arranged in showy sprays (6–8 inches long) at branch tips; **resemble lily-of-the-valley flowers. Fruits mature in autumn as small (to 1/2 inch), dry, erect capsules.** Seeds brown, pointed, 1/8-inch.

Habitat: Dry ridges and slopes near or on sandstone outcrops, especially on acid soils.

Range: Appalachian species that barely reaches Indiana at a few sites in Ohio River border counties. Sometimes planted ornamentally in Indiana.

Comments: Red-brown wood is heavy, hard, and close-grained. Used occasionally for paneling or tool handles. Flowers eagerly sought by bees provide a locally available gourmet honey.

PERSIMMON

Diospyros virginiana L.

Distinguishing Features: Medium-sized tree (to 60–80 feet tall, 1–2 feet diam.) with spreading crown; **branches often nearly at right angles to the trunk. Bark broken into rough square blocks.**

Leaves: Alternate, simple, borne singly on short (<1 inch) stems; blades oval to oblong (3–6 inches long, 1/2 as wide); **leaf edges smooth; dark glossy green** above, paler and sometimes woolly beneath; short point at tip, tapering or rounded base.

Bark: Dark gray, broken into squarish, rough, alligator-like blocks on older trees. **Young trees often show orange between the blocks.**

Twigs/Buds: Twigs slender, brown/gray, smooth to hairy, sometimes warty. Buds small (1/8 inch), rounded, smooth, reddish-brown to brown, with two dark brown scales.

Flowers/Fruit/Seeds: Flowers unisexual on separate trees. Tube-shaped male flowers (to 1/2 inch) in clusters of 2–3; solitary bell-shaped female flowers (to 3/4 inch) are cream to yellow. **Fruit a large (to 2 inches) round fleshy berry, usually orange,** sweet, and tasty when ripe, but astringent if eaten green. Contains several flat oval (to 1/2 inch) brown seeds.

Habitat: Relatively common in clearings, fencerows, and old fields, often forming thickets of small trees. Occasional in mature forests of southern Indiana.

Range: Native to southern and southwestern Indiana; widely planted or escaped from cultivation elsewhere in the state.

Comments: Wood hard, heavy, strong, and close-grained. Pretty wood of limited use in furniture manufacture; once used for shoe lasts, weaving shuttles, golf club heads, and billiard cues. Easily confused with Black gum when not fruiting—compare characteristics of the two species.

WILD BLACK CHERRY

Prunus serotina Ehrh.

Distinguishing Features: Trees of medium to large size (to 90 feet tall, 3 feet diam.) with a narrow, oblong to broadly rounded crown. **Twigs have a bitter almond taste. Leaves long, slender, finely toothed; pair of glands on leaf stem. Flowers white in long drooping clusters. Fruit clusters of purple-black cherries. Bark black, scaly.**

Leaves: Alternate on twig, simple, borne on short (<1 inch), slender, smooth leafstalks, with a pair of reddish glands near the base of the blade. Blades to 5 inches long, 1/3 as broad; edges finely and blunt-toothed. Dark glossy green above; pale and smooth below, except for rows of orange-rusty hairs along the main vein. Turn bright yellow in autumn.

Bark: Smooth, thin (sometimes peeling), red-brown, with **long whitish "pores" on young stems.** Becoming black and breaking into hard scaly plates on older trees, resembling burnt corn flakes.

Twigs/Buds: Twigs slender, smooth, red-brown, with abundant elongated pores; have bitter taste when chewed. Leaf scars half-round. Buds oval, sharp-pointed, smooth, to 1/4 inch long; with dark-brown overlapping scales.

Flowers/Fruit/Seeds: Flowers bisexual (to 1/4 inch diam.), with five white petals; crowded in showy, drooping clusters (to 6 inches long), after leaves emerge. Fruits are clusters of purple-black, juicy cherries (to 1/3 inch diam.). One oval (to 1/4 inch) stone in each edible, although slightly bitter, fruit. Scattered by birds and small mammals.

Habitat: Moist, fertile soil of upland woods, fencerows, roadsides, forest margins. Requires open sun for optimum seedling establishment.

Range: Throughout Indiana.

Comments: **Light reddish-brown wood has a beautiful smooth grain; hard and strong. Works beautifully into cabinets, furniture, or interior finishing. Wood also used for cutlery handles and printer's blocks.** Makes an attractive ornamental. Wilted leaves and twigs contain cyanic acid, which can be fatal to livestock if eaten in quantity.

PIN-CHERRY; FIRE CHERRY

Prunus pensylvanica L.f.

Distinguishing Features: Trees small (usually <30 feet tall, to 1 foot diam.) with a short trunk and narrow crown of slender, horizontal branches. **Flowers in flat-topped clusters. Fruit a small red cherry with a single stone.**

Leaves: Alternate on twig, simple, borne on slender, smooth leafstalk (to 1 inch long), often with glands at base of the blade. Blades oval to lance-shaped, tapering to a long-pointed tip, base rounded; **edges sharply and finely toothed with incurving teeth.** Bright green and lustrous above, paler and smooth below. Turn bright yellow to orange-red in autumn.

Bark: Thin, red-brown, aromatic, smooth to scaly, sometimes peeling into papery sheets; usually marked by bands of orange "pores."

Twigs/Buds: **Twigs slender, bright red to wine-colored, smooth, shiny; often with spine-like side branches.** Leaf scars alternate, half round. Buds small (<1/8 inch long), oval, pointed, with bright reddish-brown scales.

Flowers/Fruit/Seeds: Flowers bisexual, occur with the leaves; to 1/3-inch diameter in flat-topped clusters of 4–5; 5 white petals. Fruit red, round (to 1/4 inch diam.), on stems to 3/4 inch, single oval stone, flesh sour. Seed is bird-scattered.

Habitat: **On dry ridges with black oak; favored by burning.**

Range: **Few county areas of northwestern and far northern Indiana.**

Comments: Light brown, soft, close-grained; with thin yellow sapwood. Trees too small for use as a wood source. Grows rapidly; short-lived successional species; often invades after fires.

AMERICAN WILD PLUM

Prunus americana Marshall.

Distinguishing Features: Small spiny-branched tree (to 25 feet tall, 1/2 foot diam.) with a broad, irregular crown of crooked branches. Usually spreads from root sprouts into dense thickets. **Flowers in showy white clusters. Fruits fleshy, red plums to 1-inch diameter. No glands on teeth of leaves.**

Leaves: Alternate on twig, simple; leafstalks slender, usually smooth, short (<1 inch). Blades oval (to 4 inches long, 1/2 as wide), broadest at middle, the tip long-pointed, base rounded, edges finely toothed—the teeth *not* glandular. Dark green above, paler below; both surfaces often hairy.

Bark: Thin, dark reddish-brown, and smooth when young. Dark brown, rough, and scaly with age.

Twigs/Buds: Twigs slender, shiny, dark red-brown, usually with short thorn-like side branches and numerous "pores." Leaf scars alternate, half round, elevated. Buds oval (<1/4 inch long), pointed, covered with red-brown overlapping scales.

Flowers/Fruit/Seeds: Flowers bisexual, showy, several in a cluster, with 5 white petals (flowers to 1 inch across). **Thickets of trees often snowy white before leafing.** Fruits globe-shaped, large (to 1 inch diam.), red or with whitish waxy bloom when ripe; flesh bright yellow, juicy, sour but edible (good for jam). Single round, hard, flattened stone (seed), to 3/4 inch across.

Habitat: Fencerows, old fields, pastures, stream banks, woods margins—**often in dense thickets.**

Range: Throughout most of Indiana.

Comments: Reddish-brown wood is hard, strong, and close-grained, but little used due to small tree size. Good for turning on lathe. Fruit useful for jelly, jam, preserves, and pies. Valuable species for wildlife food and cover.

218

HORTULAN PLUM

Prunus hortulana L. H. Bailey.

Distinguishing Features: Trees small (to 20 feet tall, 1/2 foot diam.) with broad, rounded crown; usually does not form thickets. **White flowers appear after leaves emerge. Leafstalks glandular. Buds round-tipped. Fruit pits pointed at each end.**

Leaves: Alternate on twig, simple, borne on slender, hairy, glandular leafstalks (to 1 inch long). Blades oval to oblong (to 6 inches long, 1/3 as broad), broadest near the tapered base, tip long-pointed, edges finely toothed. Dark green and smooth above, paler and often hairy below.

Bark: Gray to brown, scaly at trunk base at maturity.

Twigs/Buds: Twigs slender, reddish-brown, smooth. Leaf scars alternate, half-round, elevated. Buds oval (to 1/4 inch long), round-tipped, smooth, with red-brown scales.

Flowers/Fruit/Seeds: Flowers bisexual, showy, several in a cluster, with 5 white petals (flower to 3/4 inch across), appear after leaves are partly expanded. **Fruits globe-shaped (to 1 inch diam.), dark red to yellow, flesh edible, but sour.** Stone compressed, pointed at each end.

Habitat: Well-drained soil at woods edges, in fencerows, and successional old fields.

Range: Scattered in southern two-thirds of Indiana, mostly in southwestern part of the state.

Comments: Wood reddish-brown, hard, close-grained, and heavy, but of little use due to small tree size. Fruit occasionally used for jam or preserve, but not as tasty as other wild plums. (***Prunus munsoniana*** [also called Wild goose plum] is grouped with *P. hortulana* for our purposes here, although the former species is more commonly found on wet sites and also forms dense thickets.)

CANADA-PLUM

Prunus nigra Aiton.

Distinguishing Features: A small tree (to 25 feet tall; <1 foot diam.) **Leaves are oval,** 2–4 inches long, half as wide, **toothed edges,** somewhat hairy below. Borne alternate on short (<1 inch) leafstalks. Flowers white to pale pink, large (to 1 1/2 inches), showy, in small clusters. **Fruit is orange-red, about 1 inch, yellow-fleshed, with a single oval stone;** good for eating, jams, or jellies. **Bark gray-brown; peels off in layers.**

Comments: Found sparingly in Indiana, apparently largely near Lake Michigan and in northern counties. Wood heavy, hard, but trees too small to be of value. Makes a nice ornamental or wildlife species.

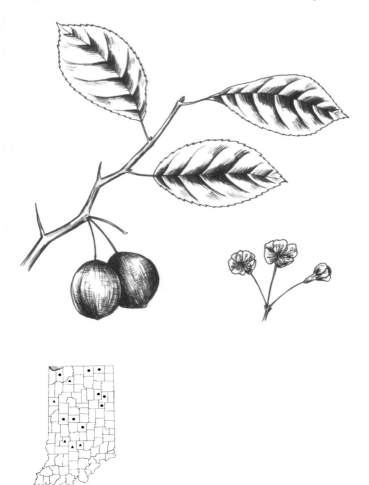

SWEET CRABAPPLE; AMERICAN CRABAPPLE

Pyrus coronaria L.

Distinguishing Features: Trees are small (to 25 feet tall, <1 foot diam.) with a broad, bushy crown. **Flowers with five white to pinkish petals, showy** (to 1 1/2 inches across). **Fruit a yellow-green apple to 1 1/2 inches in diameter.**

Leaves: Alternate on twig, simple, borne on stout, sometimes hairy leafstalks to 2 inches long. Blades oval (to 3 inches long, 2/3 as broad), widest near base or at middle, tip short-pointed, base rounded to heart-shaped; edges sharply and coarsely toothed, sometimes lobed. Bright green above, paler below; both sides smooth. Turn yellow to orange-red in autumn.

Bark: Gray-brown to red-brown; furrows between long scaly ridges.

Twigs/Buds: **Twigs slender, often with thorn-like spurs,** red-brown, smooth when mature. Leaf scars alternate, narrow-curved. Buds to 1/4 inch long, rounded, smooth, overlapping scales red.

Flowers/Fruit/Seeds: Bisexual flowers open after leafing, showy (to 1 1/2 inches across), clustered on slender stalks, five white to rose-colored petals. Fruit an apple to 1 1/2 inches across, wider than tall; pale green, fragrant, with waxy surface; edible but very tart. Seeds dark brown, flattened.

Habitat: Roadsides, fencerows, along stream banks, open woods, and clearings.

Range: Scattered throughout most of Indiana.

Comments: Wood heavy, soft, and close-grained; red-brown with thick yellow sapwood. Used for firewood, tool handles, turning stock. Fruit can be used for jelly and pies. Makes a nice ornamental tree.

PRAIRIE CRABAPPLE

Pyrus ioensis (A. Wood) L. Bailey.

Distinguishing Features: Trees small (to 25 feet tall, <1 foot diam.) with open, spreading crowns. Frequently grows in dense thickets. **Leaves irregularly toothed and lobed. Flowers large (to 2 inches across), showy and *hairy*. Fruit a yellow-green apple.**

Leaves: Alternate on twigs, simple, borne on stout, hairy leafstalks (to 1 1/2 inches). Blades oval (to 4 inches long, <1/2 as broad); rounded to pointed tip, base rounded; edges irregularly toothed and shallowly lobed. Dark green, smooth above; paler and hairy below. Turn yellow to orange-red in autumn.

Bark: Red-brown with lengthwise furrows and scaly ridges.

Twigs/Buds: **Twigs slender, reddish-brown; some with sharp thorns on side branches.** Leaf scars alternate, narrow, curved. Buds small (1/8 inch), oval, with hairy, reddish-brown scales.

Flowers/Fruit/Seeds: Flowers bisexual, large (to 2 inches across), in showy clusters, densely hairy; with five rounded, white to rose-colored petals. Fruits are nearly round, yellow-green apples to 1 1/4 inches diameter, depressed above and below. Edible, though somewhat bitter; can be used for jelly, cider, and vinegar. Seeds flattened (to 1/4 inch), dark brown.

Habitat: Open woods, fencerows, stream banks, and edges of fields and prairies.

Range: Scattered, largely in western half of Indiana.

Comments: Brown wood is heavy, close-grained. Trees used for firewood or small wooden items, but are too small for any commercial importance. A pretty ornamental and sometimes used as such.

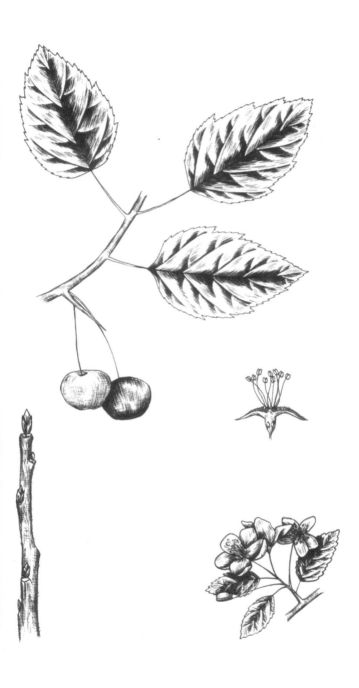

COCKSPUR-THORN

Crataegus crus-galli L.

Distinguishing Features: Trees small (to 20 feet tall, <1 foot diam.), with widely spreading branches. Leaves are leathery, dark green, shiny, and widest above the middle. **Twigs (and sometimes the trunk) armed with long, slender, sharp thorns (to 3 inches), which resemble a cock's spur** (hence common name). Flower stalks smooth. Fruits small (to 1/2 inch diam.), green to dull red.

Leaves: Alternate on twig, simple, borne on stout, smooth leafstalks (to 1 inch long). Blades oval, broadest above the middle; tip short-pointed, tapering to the base; to 2 1/2 inches long, <1/2 as wide; edges sharply toothed, except near the base. **Dark green and shiny above, paler and smooth below. Leaves leathery in texture.** Turn yellow in autumn.

Bark: Dark brown, scaly.

Twigs/Buds: Twigs moderately stout, smooth, somewhat zigzag, light brown, with long sharp thorns. Leaf scars alternate; elevated, crescent-shaped. **Buds** rounded (to 1/4 inch diam.), smooth, **red to red-brown.**

Flowers/Fruit/Seeds: Flowers bisexual, showy (to 2/3 inch across), white, several to a cluster, 5 petals. **Fruits nearly round, green to dull red, fleshy but dry, not edible; resemble small apples;** contain 1–3 nutlets. Bird- and mammal-scattered.

Habitat: Thickets, pastures, woods borders, wooded slopes.

Range: Scattered across Indiana. Many gaps in county records; needs field verification of occurrence.

Comments: Wood heavy, hard, close-grained, heartwood brown; little used due to small tree size. Good for firewood. Sometimes used as an ornamental.

GREEN HAWTHORN

Crataegus viridis L.

Distinguishing Features: Small tree (to 30 feet tall, <1 foot diam.) with a spreading crown of slender branches. **Leaves oval, widest at middle, edges toothed to lobed, largely hairless. Thorns absent or few and scattered** (to 1 1/2 inches). Flowers small (to 2/3 inch diam.); fruits small (to 1/3 inch diam.), bright red to orange red.

Leaves: Alternate on twig, simple, borne on slender leafstalks (to 1 inch long). Blades thin, oval (to 2 1/2 inches long, 1/2 as broad), widest near the middle; tip short-pointed, tapered to base; edges finely toothed to lobed near tip; veins impressed. Yellow green, smooth above; paler with hair tufts in vein angles below. Turn yellow in autumn.

Bark: Pale gray outside, thin, scaly; inner bark orange-brown.

Twigs/Buds: Twigs slender, often thornless, pale gray. Leaf scars alternate, crescent-shaped. **Buds rounded, red to red-brown.**

Flowers/Fruit/Seeds: Flowers bisexual, showy, creamy white (sometimes pink), in clusters, to 2/3 inch across, petals 5. Fruits nearly round, small (to 1/3 inch), red to orange-red, juicy. Nutlets usually 5. Most commonly bird-scattered.

Habitat: **Usually in low-wet woods or along stream banks.**

Range: **Few counties in far southwestern Indiana.**

Comments: Wood properties and uses similar to other hawthorns. Little used due to small tree size. A largely thornless tree; when thorns are present they are usually less than 1 1/2 inches long.

DOTTED HAWTHORN

Crataegus punctata Jacq.

Distinguishing Features: Trees small (to 25 feet tall, <1 foot diam.) with open, spreading crowns and many stout thorns on its twigs and branches. **Leaves oval, broadest above the middle; edges toothed to shallowly lobed. Fruits** (to 1 inch diam.) dull red (sometimes bright yellow), and **covered with many pale dots.**

Leaves: Alternate on twig, simple, borne on slender, hairy leafstalks (to 1 1/2 inches). Blades are thick, ovate, broadest above the middle (to 2 1/2 inches long, 2/3 as wide); tip short-pointed, tapering to base; edges singly (sometimes doubly) toothed above middle of leaf, occasionally lobed. Gray-green above, paler and hairy below. Turn yellow in autumn.

Bark: Gray-brown, becoming furrowed and scaly with age. **Often with long (to 3–4 inches) branched thorns on trunk.**

Twigs/Buds: **Twigs pale gray, armed with slender spines (to 2 1/2 inches long).** Leaf scars alternate, crescent-shaped. **Buds** small, rounded, with **red to red-brown** scales.

Flowers/Fruit/Seeds: Flowers bisexual, showy (to 3/4 inch diam.), white with 5 petals; in many flowered clusters. Fruits globe-shaped (to 3/4 inch diam.), dull red to bright yellow, covered with many pale dots (hence common name). Nutlets 3–5. Bird- or mammal-scattered.

Habitat: Woods borders, abandoned pasture fields, rocky slopes—often forming thickets.

Range: Scattered across eastern half of Indiana. Records few for western part of the state.

Comments: Wood properties and uses similar to those of other hawthorn species. Use limited by small size. May be used ornamentally. Formerly used by shrikes as a nest tree, and the thorns for impaling and storing prey.

DOWNY HAWTHORN

Crataegus mollis (T. & G.) Scheele.

Distinguishing Features: Small trees (to 30 feet tall, 1 foot diam.), with widely spreading crowns. **Twigs and small branches rarely armed with thorns. Ovate leaves are usually densely hairy on both surfaces. White flowers are hairy. Fruit bright red, large (to 1 inch diam.), edible.**

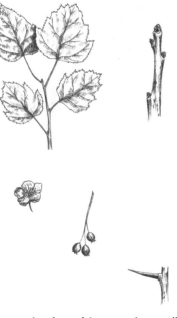

Leaves: Alternate on twig, simple, borne on stout, hairy leafstalks (to 1 inch long). Blades oval, broadest near the base (to 4 inches long, almost as broad); short-pointed at tip, base rounded to heart-shaped; **coarse-toothed to shallowly lobed edges.** Yellow-green above, paler below; both surfaces usually hairy. Turn bright yellow to yellow-orange in autumn.

Bark: Gray to brown, scaly; deeply furrowed on older trees.

Twigs/Buds: Twigs moderately stout, gray to brown, usually smooth; rarely with thorns. Leaf scars alternate, small, crescent-shaped, elevated. **Buds** small (<1/4 inch diam.), nearly round, **red to red-brown** scales, usually hairy.

Flowers/Fruit/Seeds: Flowers bisexual, open April–May; showy (to 1 inch across), several in a cluster, 5 white petals. Fruits nearly round, large (to 1 inch); bright red, look like small apples; fleshy but dry, edible; contain 4–6 hard nutlets. Bird- and mammal-scattered.

Habitat: Moist woods, sometimes in fencerows or pastures.

Range: Scattered throughout much of Indiana; apparently rare in southwestern counties.

Comments: Wood heavy, hard, close-grained; heartwood brown. Small tree size limits use to tool handles, canes, and lathe-turned items. All hawthorns can be used as ornamentals, and all have wildlife value for food and cover. Also called Red haw.

234

DOWNY SERVICEBERRY

Amelanchier arborea (Michx. f.) Fern.

Distinguishing Features: Trees are small (to 30 feet tall, <1 foot diam.), with open spreading crown of slender branches; trunk often divided into several stems. **Distinctive delicate white flowers before leafing. Fruits in clusters, berry-like, dark red.**

Leaves: **Alternate on twig, simple,** borne on slender, hairy leafstalks (to 1 1/2 inches). **Blades oval** (to 3 inches long, 1/2 as wide), **edges finely and sharply toothed,** tip short-pointed, base rounded or heart-shaped. Dark green and smooth above, paler and usually silver-haired below. Turn a beautiful deep yellow to orange-red in autumn.

Bark: **Smooth and silvery gray** when young; becomes nearly black and breaks into scaly ridges with age.

Twigs/Buds: Twigs slender, smooth, gray to red-brown, bitter to taste. Leaf scars alternate, narrow crescent-shaped. **Buds long and slender (to 1/2 inch), pointed; with reddish-brown overlapping scales.**

Flowers/Fruit/Seeds: Delicate white flowers appear in drooping clusters before leafing; bisexual, 5 strap-like petals are 1/2+ inch long, often twisted. **Fruits are reddish-purple, round (to 1/3 inch), fleshy berries;** very tasty, eagerly sought by birds and other wildlife; contain 2–5 seeds that have an almond taste.

Habitat: Upland sites, usually with fertile soils—forest edges, wooded slopes, sometimes at bluff edges or in fencerows.

Range: Scattered throughout much of Indiana; apparently less commonly found in southwestern and central parts of the state.

Comments: Wood brown to dark red-brown, very hard, heavy, and close-grained; used for tool handles, turned articles, and firewood, but trees are too small to harvest. An excellent and widely planted ornamental as it is beautiful in all seasons. Fruit excellent for pies, muffins, and preserves if you can beat the wildlife to the berries. Also called Juneberry and Shadbush.

SMOOTH SERVICEBERRY; ALLEGHENY SHADBUSH

Amelanchier laevis Wieg.

Distinguishing Features: Trees small (to 20 feet tall, <1 foot diam.), with a narrow rounded crown of slender, spreading branches. **Leaves not hairy;** bronze-purple through flowering time. Leafstalks and flowerstalks without hair. Fruits purple-black, juicy, sweet, and edible.

Leaves: **Alternate on twig, simple,** borne on slender, hairless leafstalks (to 1 inch long). **Blades oval** (to 3 inches long, 1/2 as broad); tip pointed, base rounded; **edges finely toothed.** Dark green above; paler below; **both surfaces without hair.**

Bark: **Smooth and pale gray** at first; darker and scaly on mature trees.

Twigs/Buds: Twigs slender, smooth, gray to reddish-brown. Leaf scars alternate, elevated, narrowly crescent-shaped. Buds slender, pointed (to 3/4 inch long), smooth, with overlapping red-brown scales.

Flowers/Fruit/Seeds: Flowers bisexual, appear before leaves are full-sized. Showy, large, in drooping clusters; 5 long (to 1 inch), narrow, strap-like petals; flowers smooth. **Fruits are dark purple-black berries (to 1/3 inch diam.),** fleshy, sweet, juicy, edible. Have two to several almond-flavored seeds.

Habitat: Moist woods, slopes, rural roadsides, forest borders.

Range: Scattered throughout much of the state. Apparently rare to absent in southwestern and central Indiana.

Comments: Wood characteristics and uses the same as for the very similar Downy serviceberry (which see). These species were called serviceberry by the pioneers (or "sarvis" in the South) because they flowered at the time of the first church service given by circuit-rider preachers. This species also called Smooth shadbush (for flowering at the time shad swim upstream to spawn) or Allegheny juneberry.

REDBUD

Cercis canadensis L.

Distinguishing Features: Small tree (usually <30 feet tall, <1 foot diam.) with a broad, spreading crown; often with multiple trunks. **Beautiful pink-red flowers in spring. Leaves heart-shaped with smooth edges.**

Leaves: Alternate, simple, borne singly on long (to 4 inches) leafstalks; **leaf stem is enlarged and fleshy for 1/4 inch where it joins blade.** Blades broad heart-shaped (to 5 inches long), edges smooth, dull point at tip; both surfaces smooth at maturity. Bright yellow in autumn.

Bark: Dark gray, thin, becoming scaly with age; underbark red-brown in fissures.

Twigs/Buds: Twigs slender, brown, warty, often zigzag; leaf scars triangular, elevated. End buds absent; side buds small, chestnut-brown, scaly.

Flowers/Fruit/Seeds: Pea-shaped (to 1/2 inch long), **bisexual flowers,** which appear in clusters before leafing in April/May, **are beautiful rose-pink to purple-red.** Fruit a flattened, 3- to several-seeded pod (to 3 inches long, 1/2 inch wide); brown when ripe in August. Seeds flattened, small (to 1/4 inch), brown, very hard.

Habitat: Rich, moist woods, roadsides, and woodland borders. Widely planted as a showy, much-loved ornamental.

Range: Throughout Indiana.

Comments: Wood hard, heavy, durable; heartwood dark yellow-brown; good firewood, but of no economic value. A white-flowered form is occasionally found. Sometimes called Judas tree.

HONEY-LOCUST

Gleditsia triacanthos L.

Distinguishing Features: Medium to large trees (to 80 feet tall, 3 feet diam.) with open lacy crowns. **Leaves mostly twice-compound with numerous small leaflets. Usually with long (to 12 inches), 3-parted (hence species name), purplish-brown thorns. Fruits are long, strap-like pods.**

Leaves: Alternate, mostly twice compound, to 10 inches long; leaflets many, small (to 1 inch), oblong, edges smooth to wavy, stem fuzzy, tips blunt, dark green surfaces. Turn yellow in autumn.

Bark: Dark gray, scaly; scales seem to curl from trunk along one edge. Trunks usually have long, branched thorns.

Twigs/Buds: Twigs somewhat stout, angular, reddish-brown, smooth, zigzag, thorns present. **Buds small (1/8 inch), rounded, hairy, nearly sunken in twig.**

Flowers/Fruit/Seeds: Unisexual, male and female usually on separate trees; appear after leafing. Male flowers in dense clusters; female in slender clusters of few flowers. **Fruits are long (to 18 inches), flattened, purple-brown pods** (legumes); contain few to several flattened, chestnut brown seeds (to 1/3 inch long).

Habitat: Moist wooded ravines and bottomlands along streams. Sometimes in fencerows and old pasture fields.

Range: Throughout Indiana.

Comments: Flowers have abundant nectar and pulp of fruit is sweet—hence the common name. A thornless variety (Moraine locust) is a very popular and widely planted ornamental. Wood hard, strong, durable, a pretty pale red-brown. Used for fence posts, firewood, rough construction.

WATER-LOCUST

Gleditsia aquatica Marshall.

Distinguishing Features: Medium-sized tree (to 60 feet tall, 2 feet diam.) with an irregular, spreading crown. **Leaves singly or twice compound. Fruit a short pod with 1–2 seeds, but no pulp. Thorns on trunk are branched; those on limbs flattened.**

Leaves: Alternate, both compound and twice compound on same tree; leaflets oval-oblong (to 1 inch long, 1/2 as wide), blunt tips, edges smooth to ragged, surfaces largely hairless.

Bark: Dark gray to dark brown, smooth to furrowed with age. Branched thorns usually present on trunk.

Twigs/Buds: Twigs slender, smooth, greenish to gray; usually with flattened unbranched thorns. Buds small (to 1/8 inch), dark brown, nearly hidden by leaf scars.

Flowers/Fruit/Seeds: Some flowers bisexual, others unisexual, in greenish long (to 4 inches) clusters following leafing in May/June; individual flowers small, sweet pea–like. **Fruit short, flat** (1–2 inches, 1 inch wide) **pods, pointed at both ends;** without pulp; in drooping clusters; contain 1–2 round, flattened, brown seeds to 1/2 inch diam.

Habitat: Low swampy woods in river sloughs, with roots frequently submerged for several months.

Range: **Known only from Knox, Gibson, and Posey Counties** in southwestern Indiana.

Comments: Wood heavy, durable, coarse-grained, heartwood reddish-brown; of no economic importance here. A rare, locally restricted tree in Indiana.

KENTUCKY COFFEE-TREE

Gymnocladus dioica (L.) K. Koch.

Distinguishing Features: Medium-sized trees (to 100 feet tall, 2+ feet diam.) with open crowns. **Leaves twice compound; leaflets large. Branches very stout, appear naked (hence generic name of tree). Bark dark gray, scaly. Fruits are short, thick, hard pods.**

Leaves: Alternate, twice compound, very large (to 36 inches long, 1/2 as broad); leaflets usually alternate, dark green, numerous, 1–1 1/2 inches long, oval, edges smooth, tip pointed, hairy at first, smooth at maturity.

Bark: Dark gray, furrowed, the ridges and scales often curling along their sides; scales hard and sharp-edged.

Twigs/Buds: **Twigs stout, crooked, gray-brown; leaf scars large, heart-shaped; pith large, salmon-colored. Buds tiny, sunken in hairy cavities above leaf scars (superposed).**

Flowers/Fruit/Seeds: Unisexual with male and female on separate trees after leaves emerge. Clusters of male flowers to 4 inches long; female flower clusters to 12 inches. **Fruits are thick, tough, hard, purple-brown pods** (legumes) to 6 inches long; **contain several very hard, large** (to 5/8 inch) **seeds in a sticky, greenish pulp.**

Habitat: Rich moist woods, often on bottomlands. Occasionally in fencerows or lawns.

Range: Essentially throughout Indiana, but rather widely scattered.

Comments: Wood heavy, strong, coarse-grained, pink alternating with brown. Used for fence posts and firewood, but is a pretty cabinet wood. Pioneers roasted seeds as a coffee substitute, hence the common name. Sometimes root sprouts into rather large colonies.

YELLOW-WOOD

Cladrastis lutea (Michx. f.) K. Koch.

Distinguishing Features: Medium-sized tree (to 50 feet tall, <18 inches diam.); crowns broad, rounded. **Leaves once compound with smooth edges. Smooth gray bark resembles that of American beech. Flowers in showy white clusters; fruit a flattened pod.**

Leaves: Alternate, ladder-like (pinnately) compound, 8–12 inches long; oval leaflets usually 7–11, tips dull-pointed, tapering to base, edges smooth; yellow-green above, paler below, blades hairless. **Leaflets arranged alternately on central leaf stem. Leafstalk base swollen and hollow; covers bud.** Turn yellow in autumn.

Bark: Thin, smooth, silvery gray; similar to that of American beech.

Twigs/Buds: Twigs slender, brownish, somewhat zigzag, brittle; leaf scars horseshoe-shaped, encircle small (1/8 inch) hairy buds which lack scales.

Flowers/Fruit/Seeds: Bisexual white flowers borne in long (to 1 foot), drooping clusters, during June; mildly fragrant, very showy. Individual flowers sweet pea–like. Fruits long (2–4 inches), flattened, smooth brown pods (legumes), each containing 4–6 small, flat seeds. Fruits mature August/September.

Habitat: Rich, moist, wooded slopes and deep ravines.

Range: Occurs naturally at a few sites and only in Brown County. A lovely ornamental found occasionally in lawns and parks.

Comments: Wood hard, heavy, strong, close-grained, yellow (hence common name). Once favored for gunstocks by pioneers. Generic name from the Greek; means "brittle branches."

BLACK LOCUST

Robinia pseudoacacia L.

Distinguishing Features: Medium-sized tree (to 75 feet tall, 2 1/2 feet diam.) with narrow, oblong, open crown of spreading branches. Ladder-like **once-compound leaves with smooth edges, and paired short spines near leaf base. Flowers in showy white clusters; fruit a narrow, brown pod.**

Leaves: Alternate, once compound with 7–19 oval leaflets on short stems; leaflet margins entire, bases and tips round; dark green above, paler and somewhat hairy below, especially on veins. Turn yellow in autumn.

Bark: Light brown with many warty "pores" on young trees. Dark gray to black, deeply furrowed with elevated fibrous ridges when mature. **Inner bark and bark of roots yellow to orange.**

Twigs/Buds: Twigs slender, zigzag, brown; often with paired spines at side buds. Buds small (to 1/8 inch), embedded in twig; without scales.

Flowers/Fruit/Seeds: Bisexual flowers pea-like (to 1 inch), white with a yellow spot, in long drooping clusters, bloom May/June, very fragrant. **Fruits are narrow** (to 1/2 inch), **brown, straplike pods** (legumes) to 4 inches long; each with 4–8 flat, kidney-shaped seeds.

Habitat: Drier woodlands, thickets, roadsides, old fields. Widely planted from earliest pioneer days to present time. Also used as an ornamental, but often escapes and becomes invasive.

Range: Apparently native only in southeastern Indiana near the Ohio River. Now occurs throughout the state.

Comments: Hard, heavy, strong, very durable in soil; used for fence posts, mine timbers, railroad ties, and insulator pins. Commonly planted for surface mine reclamation. Flower an important nectar source for honeybees.

250

FLOWERING DOGWOOD

Cornus florida L.

Distinguishing Features: Small tree (to 30 feet tall, <1 foot diam.) with a spreading, rounded crown. **Beautiful large white "flowers" in spring. Opposite, toothless leaves with impressed, curved veins.** Bark of older trees broken into rounded scales.

Leaves: Opposite, simple, borne on very short (1/2 inch) leafstalks. Blades oval, 2–5 inches long, pointed at tip, bases taper to stem, veins impressed and curved upward, leaf edges smooth. Bright green above, paler below, little hairiness. Turn a beautiful red in autumn.

Bark: Rough, dark gray to dark brown, broken into small, **thin rounded scales that resemble coins.**

Twigs/Buds: Twigs slender, green to purple, often white-frosted; tips curve upward. Leaf scars opposite, crescent-shaped, elevated. Buds of two kinds—leaf buds slender, pointed; flower buds chubby, top-shaped.

Flowers/Fruit/Seeds: **Bisexual flowers** open before leaves in late April–May; several in dense, greenish-white clusters, each **surrounded by four large, showy, petal-like, white bracts. Fruits ripen September/ October; clusters of bright scarlet oval shiny berries** (to 1/2 inch); 1–2 seeds—oval, pointed pits—in each fruit.

Habitat: Upland woods, roadsides, woody old fields. Beautiful ornamental and widely planted.

Range: Throughout Indiana.

Comments: Wood heavy, hard, tough, and fine-grained. Once used for weaving shuttles, pulleys, mallets, golf club heads, and thread spools. Slow-growing tree, but widely used in landscaping. Fruit important to wildlife. Natural stands suffering great mortality due to powdery mildew and anthracnose.

BLACK GUM

Nyssa sylvatica Marshall.

Distinguishing Features: Medium-sized tree (to 90 feet tall, 2–3 feet diam.) with spreading crowns. **Branches nearly at right angles to trunk in younger trees. Bark in alligator-like blocks on older trees.** Trunk bases of older trees often swollen and hollow. Easily confused with persimmon due to similarities in leaves and bark.

Leaves: **Alternate, simple, borne singly on short (<1 inch) stems, but clustered near twig ends. Blades egg-shaped, widest above middle** (to 6 inches long; 1/2 as wide); **edges smooth,** short-pointed at tip, tapering at base; shiny dark green above, paler and somewhat hairy below. Turn beautiful deep red to purple in autumn.

Bark: Bark is dark brown to gray-black; deeply fissured on older trees and breaks into blocks that resemble alligator hide.

Twigs/Buds: Twigs smooth, medium diameter, reddish-brown; leaf scars crescent-shaped; pith has partitions. Buds small (to 3/16 inch), short-pointed, red-brown, with few hairy scales.

Flowers/Fruit/Seeds: Unisexual flowers usually on separate trees in May–June; male flowers small, greenish, in clusters; female, two to several on long stalks near leaf stems. **Fruits fleshy, plum-like (to 1 inch), oval, blue-black, sour,** on 1–3 inch stems; ripen October. The single seed has pointed ends and several ridges. Scattered by birds and small mammals.

Habitat: Flat wet woods and upland ridges, usually on acid soils. Occasionally in fencerows and old pastures. Beautiful ornamental and widely planted.

Range: Common in southern half of Indiana and in northern third—rare to absent on central tillplain.

Comments: Wood is strong, tough, and hard to split. Used for stable floors, planking, barrels, crating, and some furniture. Hollow sections used as beehives by pioneers—"bee gums." Also called Sour gum, Black tupelo, and Pepperidge.

OHIO-BUCKEYE

Aesculus glabra Willd.

Distinguishing Features: Medium-sized tree (to 60 feet tall, 1–2 feet diam.) with a broad, rounded crown; branches often droop. **Pale gray bark. Leaves opposite, compound, hand-shaped.** Flowers in large, showy clusters. **Distinctive "buckeye" seeds** resemble eye of deer. **Has offensive odor** when leaves, twigs, or bark are bruised— "stinking buckeye."

Leaves: Opposite, borne palmately compound with 5 or 7 finely toothed leaflets (3–6 inches long) on long (3–5 inches) leaf stems. Oval leaflets without stems, tips tapering to a point, bases likewise, essentially hairless; yellow-green above, paler below. Turn yellow in autumn.

Bark: **Ash-gray;** smooth when young; thick, plated, and fissured on older trees.

Twigs/Buds: Twigs stout, brown, prominent leaf scars are triangular. **Buds large** (to 2/3 inch), **orange-brown,** sometimes sticky, with large, keeled scales.

Flowers/Fruit/Seeds: **Flowers yellow-green, hairy, borne in erect, large (to 10 inches long) cone-shaped clusters at branch tips** in April/May. Fruits are nearly round, warty to spiny capsules (1–2 inches diam.) with a leathery husk; open to release 1–3 large, glossy, chocolate-colored seeds (called buckeyes). Seeds *toxic* to farm animals and *not edible* to humans.

Habitat: Usually rich, moist woods, well-drained floodplains, fencerows, and roadsides. Rarely a canopy tree in forest stands. Occasionally planted as ornamental.

Range: Throughout Indiana.

Comments: Wood lightweight, soft, nearly white, not durable. Uses: carving blocks, wooden utensils; once used for artificial limbs. In years past, superstitious people carried buckeyes in their pockets to ward off rheumatism, or just for good luck.

YELLOW BUCKEYE

Aesculus flava Aiton.

Distinguishing Features: Medium- to large-sized trees (to 80 feet tall, 2–3 feet diam.) with broad rounded crowns; gray, scaly bark. **Leaves opposite, compound, hand-shaped.** Flowers in large, showy clusters. **Distinctive "buckeye" seeds** resemble eye of deer.

Leaves: Opposite, borne palmately compound with 5 or 7 sharply toothed leaflets (each 4–6 inches long) on long (4–6 inches) leafstalks. Leaflets with very short (<1/2 inch) stems, tips tapering to a point, bases likewise, often hairy; dark yellow-green above, paler below. Turn clear yellow in fall.

Bark: Gray to dark brown and scaly; inner bark bright yellow to red.

Twigs/Buds: Twigs stout, smooth, orange-brown to brown. **Buds large (to 3/4 inch), not sticky,** covered with broad, slightly fuzzy pale brown scales.

Flowers/Fruit/Seeds: Flowers yellow, hairy, borne in large, branched clusters at branch tips; blooms April/ May. **Fruits are leathery,** 3-parted capsules 2–3 inches long; **husks thin, smooth, brownish.** Seeds large (to 1 1/2 inches), glossy brown with conspicuous pale scar; usually 2 per capsule. Seeds contain some *toxic* material; *not edible for humans,* but eaten by livestock.

Habitat: Moist woods with good drainage, ravines, and streambanks. Often a canopy tree due to its large size. Does not occur in old fields. Sometimes used as ornamental.

Range: Only in counties near Ohio River in southeastern Indiana.

Comments: Wood properties and uses similar to that of Ohio-buckeye. Also called Sweet buckeye as it **lacks disagreeable odor,** and seeds are less toxic than Ohio-buckeye.

SUGAR-MAPLE

Acer saccharum Marshall.

Distinguishing Features: Large, handsome tree (to 100 feet tall, 3 1/2 feet diam.) with dense, broad, rounded crown. **Leaves opposite, lobed, simple. Fruits paired double propellers.**

Leaves: Opposite on twig, simple; borne on slender leafstalks 3–4 inches long. **Lobes hand-like, 5-lobed with wide notches rounded at angles;** edges with few coarse teeth. Blades 3–6 inches across, nearly circular in outline; smooth, bright green above, paler below. Leaves turn bright yellow, orange, or red in autumn.

Bark: Smooth, gray to pale tan when young, becoming dark gray and deeply furrowed to shaggy with age.

Twigs/Buds: Twigs slender, greenish early, then shiny red-brown when older. Leaf scars opposite, U-shaped. Winter buds medium (to 3/8 inch long), pointed, smooth, with overlapping scales.

Flowers/Fruit/Seeds: Flowers soft yellow-green without petals, open before leafing; unisexual, male and female flowers on same tree; each flower hanging on long (1–3 inches) threadlike, slender stalk. **Fruits medium (to 1 1/4 inches), paired, U-shaped with parallel wings about 1 inch long;** mature in autumn; wind-scattered.

Habitat: Moist upland sites with deep, fertile soils; a major component of such old-growth forests, especially, since it is a slow-growing, long-lived species. Aggressive reproduction under forest canopies permits it to replace oaks in many stands. Its excellent shade production from very dense crowns and its beautiful fall coloration make it a prized tree for lawns, parks, and street borders.

Range: Throughout Indiana.

Comments: Wood light brown to nearly white, very hard, close-grained—valued for furniture, flooring, cabinets, charcoal, and is an excellent firewood. Bird's-eye and curly maple are rare, beautiful special wood grains. Its sap is used to make delicious maple syrup, and also maple sugar, hence its common name.

BLACK MAPLE

Acer nigrum Michx. f.

Distinguishing Features: **Very similar to sugar maple.** Medium to large trees (to 90 feet tall, 3 feet diam.). **Bark frequently darker, and its drooping leaves less lobed than sugar maple, stipules often at leaf bases.**

Leaves: Opposite on twig, simple; 3- to 5-lobed with wide notches and rounded angles; about as long (4–6 inches) as wide. **Often "goose foot"–shaped, hairy and with drooping edges.** Very dark green above (crown appears nearly black), paler below. **Leafstalk stout, long (3–5 inches), often with leafy growths (stipules) at base.** Fall color brilliant yellow to orange.

Bark: Deeply furrowed, ridged, dark gray to almost black on mature trees. Thin, smooth, and paler gray on younger trees.

Twigs/Buds: Twigs somewhat stout, orange-brown and slightly hairy. Buds medium (to 1/4 inch), egg-shaped, with hairy, overlapping scales.

Flowers/Fruit/Seeds: Flowers unisexual on same tree; tiny yellow-green; on drooping, thread-like stalks; no petals. **Fruits medium** (to 1 1/4 inches), **paired propellers,** U-shaped to spreading; mature in autumn; wind-scattered.

Habitat: Along streams and rivers on moist, but rarely flooded, sites. Also often intermixed with sugar maple in rich woodlands.

Range: Most of the northeastern two-thirds of Indiana. Scattered elsewhere.

Comments: Wood and ornamental uses similar to those of Sugar-maple. Lumber of both sold as hard maple. Also used for syrup manufacture. Hybridizes with Sugar-maple in some situations.

RED MAPLE

Acer rubrum L.

Distinguishing Features: Medium to large tree (to 90 feet tall, 3 feet diam.) with oval, spreading crown. **Leaves opposite, simple, with hand-like lobing. Red twigs without odor when bruised. Fruit a broad V-shaped double propeller.**

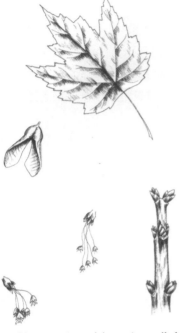

Leaves: Opposite on twig, simple, borne singly on long (to 4 inches), red stems. Lobes 3–5, hand-like, with sharp **V-shaped notches between main lobes,** usually sharply toothed. **Pale green on upper surface, gray to whitish below.** Turn scarlet or red in autumn.

Bark: Pale gray and smooth when young; becomes dark brown to gray, furrowed, then very scaly with maturity.

Twigs/Buds: Twigs slender, shiny red to reddish-brown with numerous tiny, warty bumps; leaf scars U-shaped. Buds small (<1/4 inch), rounded, usually red, with hairy scales.

Flowers/Fruit/Seeds:
Flowers unisexual, borne in small clusters on same or separate trees. Open February/March before leafing. Female flowers red; male yellow to orange. **Fruits drooping, paired, propeller-like; broadly V- or U-shaped; to 1 inch long.** Mature in spring; wind-scattered.

Habitat: Typically in wet to moist woods of bottomlands, flatwoods, and swamps, but also on acidic upland slopes and ridgetops. Invades old fields and fencerows. Very popular ornamental of lawns and parks.

Range: Throughout Indiana.

Comments: Wood heavy, close-grained, white with light brown heartwood. Sold as soft maple. Much curly maple lumber is from Red maple.
Used for furniture, flooring, cabinets, and basket veneer. Not desirable for maple syrup production.

SILVER-MAPLE

Acer saccharinum L.

Distinguishing Features: Large tree (to 100 feet tall, 4+ feet diam.) with large spreading, round-topped open crown. **Twigs red-brown with foul odor when bruised; leaves opposite, simple. Fruits are large, paired double propellers.**

Leaves: Opposite on twig, simple, borne on leafstalks 3–4 inches long. Lobes hand-like, deeply 5-lobed with **narrow notches sharp-angled,** edges sharply toothed. Blades 4–8 inches long, 2/3 as wide; **light green, smooth above, silvery white below.** Turn pale yellow in autumn.

Bark: Light gray and smooth on young stems and upper branches; becomes scaly, then shaggy with age.

Twigs/Buds: Twigs slender, green early, then rich red-brown; often curving upward at tips; disagreeable odor when bruised. Leaf scars U-shaped. Buds opposite, small (to 1/4 inch), dark red-brown with overlapping scales.

Flowers/Fruit/Seeds: Flowers in dense clusters, without petals, small, greenish-yellow to red; male and female flowers separate, but on same tree. Very early flowering. **Fruits are large (1 1/2–3 inches), paired propellers, their veined wings spreading at an angle of about 90°;** mature in spring; wind-scattered.

Habitat: Wet poorly drained lowlands, floodplains, stream and river bottoms, and along pond and lake margins. Once a very common street, park, and lawn tree; less so today.

Range: Throughout Indiana.

Comments: Wood soft, breaks and splits easily, white to pale brown, used for pulpwood, inexpensive furniture, crates—low economic value. Also called Soft maple.

BOXELDER

Acer negundo L.

Distinguishing Features: Medium-sized tree (to 60 feet tall, 3 feet diam.) with a short trunk and spreading, open crown. **Twigs green; leaves opposite, compound, with 3–5 leaflets. Fruits are paired, V-shaped double samaras.**

Leaves: Opposite on twig, compound; leaflets 3–5 (rarely 7), large, coarsely toothed to shallowly lobed. Overall leaf to 12 inches, with leafstalk 1/3 of length. Light green above, paler and usually hairy below. Turn pale yellow in autumn. **Leaves resemble those of poison ivy.**

Bark: Light brown with fine ridges when young; becomes medium gray and furrowed with age.

Twigs/Buds: **Twigs medium, green** (sometimes frosted purplish-white), smooth and shiny; leaf scars U-shaped. Buds small (to 3/16 inch), rounded, scaled, white-hairy.

Flowers/Fruit/Seeds: Flowers greenish yellow at leafing in April/May; unisexual in clusters, male and female on separate trees. Male flowers long, thread-like; female in drooping clusters. **Fruits are paired, winged seeds, which form a V-shaped double propeller (to 1 1/2 inches).** Mature in late summer; wind-scattered.

Habitat: Rich moist woods along streams and in wooded wetlands. Common in fencerows and roadsides. Once widely planted as a street or lawn tree, rarely now.

Range: Throughout Indiana.

Comments: Wood soft, weak, nearly white, often tinged with pink. Pinkish wood prized by wood turners. Used for pulpwood, boxes, crates, occasionally for furniture—low economic value. Also called Ash-leaved maple.

268

STAGHORN-SUMAC

Rhus typhina L.

Distinguishing Features: A large shrub to small tree (to 30 feet tall, 6 inches diam.) with crooked stems and scraggly, open crowns; often in dense clumps. **Leaves compound (11–25 leaflets), very large (12–18 inches), alternate.** Leaflets not stalked; edges coarsely toothed. **Stout, densely hairy twigs** (like deer antlers in velvet, hence name) **exude milky fluid when crushed.** Flowers unisexual in dense, greenish clusters at branch tips of separate plants. **Fruits in dense red clumps;** dry, hairy, small (to 1/10 inch), with hard seeds. Bark dark, gray-brown, thin; scaly when mature. Pith of stems pale brown, very large in diameter.

Comments: Widespread in fencerows, thickets, and woods edges. Plant not toxic. Wood soft, yellow-brown; has a beautiful grain; occasionally used for turnings.

WHITE ASH

Fraxinus americana L.

Distinguishing Features: Large trees (to 110 feet, 4 feet diam.) with straight columnar trunks; crown rounded with slender branches. **Leaves opposite, compound. Bark deeply fissured. Fruits single-winged, strap-like propellers.**

Leaves: Opposite on twig, compound with 5–9 (usually 7) ladder-like, oval leaflets (3–5 inches long) with pointed tips. Edges smooth to remotely toothed; smooth, dark green above, paler below. Leaf stems stout, smooth, grooved. Usually turn bright yellow in autumn.

Bark: Medium to dark gray with numerous narrow ridges crossing to form a diamond pattern. With age, it becomes very deeply furrowed with broad, scaly ridges.

Twigs/Buds: Twigs medium to stout, gray to brown with large "pores," and **opposite horseshoe-shaped leaf scars.** End buds large (to 1/2 inch), blunt, oval, with rusty brown, hairy, overlapping scales. Side buds in leaf scar notches are smaller, rounded.

Flowers/Fruit/Seeds: Flowers unisexual on separate trees; open before leafing. Both sexes without petals in purplish dense clusters, which later expand and become yellowish. **Fruits winged, paddle-shaped** (to 2 inches long), **tips sometimes notched;** several in dense, drooping clusters (6–8 inches long); mature in autumn; scattered by wind in winter or spring.

Habitat: Occurs on fertile soils along streams, on lower slopes and in moist upland forests in mixed hardwood stands. Commonly planted as an ornamental in parks and lawns.

Range: Throughout Indiana.

Comments: White to pale brown wood is lightweight, strong, and hard with a beautiful coarse grain. Used for tool handles, baseball bats, boat oars, furniture, paneling. Excellent firewood. Species of ashes are often variable and difficult to identify consistently. Also called American ash. **See twig comparison, opposite page: white-left; green-right.**

GREEN ASH

Fraxinus pennsylvanica Marshall.

Distinguishing Features: Medium trees (to 80 feet tall, 2 1/2 feet diam.), usually with straight trunks sometimes swollen at base, and rounded crowns of slender, silvery branches. **Leaves opposite, compound. Bark with diamond fissures. Fruits single-winged propellers.**

Leaves: Opposite on twig, compound, with 7–9 ladder-like, lance-shaped leaflets with remotely toothed to smooth edges. Green and hairless on both surfaces. Leaflets turn red-brown to burgundy in autumn.

Bark: Light brown-gray to medium gray with diamond-shaped furrows between flat, sometimes scaly, ridges. Occasionally lower trunk has paler-colored bark in blocky pattern.

Twigs/Buds: Twigs medium to stout, gray to brown, smooth to velvety hairy. **Leaf scars opposite, half round and straight across top.** End buds blunt, rounded (to 1/4 inch), rusty brown, scaly, and hairy. Side buds oval, smaller.

Flowers/Fruit/Seeds: Flowers unisexual and without petals in purplish clusters on separate trees; appear as leaves unfold. **Fruit** clustered on hairy stalks in autumn, a narrow, paddle-shaped one-seeded single wing (to 2 inches long); **tip not notched,** blade extends to center of seed. Wind-scattered.

Habitat: Typically in bottomland forests or moist upland woods. Very popular tree for lawns, parks, and along streets; widely planted ornamentally.

Range: Throughout Indiana.

Comments: Easily confused with white ash; their wood properties and uses are also similar. See "Species Excluded." Also called Red ash; formerly Red and Green ash were separated into different groups; some authors still separate them. Ashes are often variable and difficult to identify.

PUMPKIN-ASH

Fraxinus profunda (Bush) Bush.

Distinguishing Features: Medium to large trees (to 90 feet tall, 2 1/2 feet diam.) with a rounded open crown of stout branches. **Leaves opposite, compound. Twigs stout and velvety pubescent. Largest fruits of the ashes. Bases of trees often swollen.**

Leaves: Opposite on twigs, compound, with 7–9 leaflets which are lance-shaped with pointed tips and smooth edges. **Yellow-green and smooth above, paler and velvet-hairy below. Leaf stalks densely hairy.** Leaves turn yellow in autumn.

Bark: Light gray, fissured. Base of trees often swollen, especially in swamps.

Twigs/Buds: Twigs stout, gray to brown, usually velvety. Leaf scars opposite, half-round to horseshoe-shaped. End buds cone-shaped (to 1/4 inch), brown, and hairy. Side buds smaller, rounded.

Flowers/Fruit/Seeds: Flowers unisexual, appear in clusters on separate trees, before leafing. Small greenish-purple and without petals. Fruits, the largest of the ashes (to 3 inches long, 1/2 inch wide), are paddle-shaped with a papery wing (a single propeller). Wind-scattered.

Habitat: **Low wet woods, river sloughs, and cypress swamps, often with water standing for extended periods.**

Range: Largely in far southwestern Indiana. Widely scattered elsewhere.

Comments: Resembles White ash more than other species of ash. Wood of low value. Used for fuelwood, paper pulp, boxes, and crates. Common name presumably from the swollen, "pumpkin-like" base of trunk.

BLACK ASH

Fraxinus nigra Marshall.

Distinguishing Features: Medium trees (to 70 feet tall, 1–2 feet diam.) with narrow open crown and stout, right-angle branches. **Compound, opposite leaves with 7–11 leaflets *without* stalks. Fruit a single propeller.**

Leaves: Opposite on twig, compound, with 7–11 (commonly 9) toothed leaflets without stems. Dark green above, paler below with rusty hairs on veins. Leaf stalks stout, grooved. Leaflets turn reddish-brown in autumn.

Bark: Soft, light gray, scaly, without diamond-shaped furrows. **Flakes off when rubbed with the hand.**

Twigs/Buds: Twigs stout, dark green to gray-brown with warty "pores." Leaf scars opposite on twig, oval. End buds scaled, cone-shaped, pointed, blue-black, finely hairy (to 1/4 inch); side buds smaller, located above leaf scars.

Flowers/Fruit/Seeds: Flowers bisexual or unisexual on same tree; appear before leafing. Both sexes small, without petals, in loose clusters. **Fruits oblong (to 1 1/2 inches), winged, single propellers, notched at tip;** borne in early fall, wind-scattered in winter or spring.

Habitat: Low, cold, swampy woods around northern lakes and wetlands, and locally in certain low-ground forests of southern Indiana.

Range: Primarily in northern two-thirds of Indiana, except as noted above.

Comments: Wood is tough, coarse-grained, rather soft and dark brown (hence Black ash). Wood splits easily along annual rings; used for baskets and splints; sometimes used for cabinets, or veneer from burls.

BLUE ASH

Fraxinus quadrangulata Michx.

Distinguishing Features: Trees medium to large (to 80 feet tall, 3+ feet diam.); crowns irregular, often with dead lower branches retained at right angles to the trunk. **Leaves opposite, compound. Twigs with four ridges (sometimes with thin, narrow wings), giving a squarish appearance, hence scientific name.

Leaves: Opposite on twigs, compound, with 7–11 (usually 7) ladder-like, **spear-shaped leaflets that have very short stems, long-pointed tips, and coarse teeth on edges.** Yellow-green and smooth above, paler with hairy veins below. Turn bright yellow in autumn.

Bark: Light gray; irregularly divided into thick corky oval scales on mature trees; very shaggy on very old trees. Diamond-shaped furrows absent.

Twigs/Buds: Twigs stout, square or 4-winged, smooth. Leaf scars half round, but notched above. End buds gray, rounded, blunt (to 1/2 inch long) with fine hairs. Oval side buds smaller.

Flowers/Fruit/Seeds: Flowers are bisexual, without petals, and appear in branched, purplish clusters as leaves begin to unfold. **Fruit paddle-shaped; one-seeded; single wing (to 2 inches long); tip notched;** in clusters in autumn. Seeds wind-scattered.

Habitat: A dry, limy soil species, it is most commonly found on gravelly river bluffs, along limestone cliffs, and wooded hillsides; occasional in mesic woods of central Indiana.

Range: Scattered throughout most of Indiana except the northwest corner.

Comments: Wood is hard, heavy, and yellow-brown. Its uses are similar to those of White and Green ash. A blue dye from the inner bark of stems was used by the pioneers for coloring cloth, hence Blue ash.

280

NORTHERN CATALPA

Catalpa speciosa Warder.

Distinguishing Features: Medium to large trees (to 60 feet tall, 2+ feet diam.) with rounded crown—grew much larger in original forest. **Leaves opposite, large, heart-shaped. Bark fissured to scaly. Flowers white, showy in large clusters; fruit a long, green, round "pod."**

Leaves: **Opposite (in whorls of 3),** borne singly on long (to 6 inches), stout stems. Blades large (to 1 foot long, 2/3 as broad), heart-shaped, pointed tip, smooth edges, somewhat leathery; dark green above, paler and hairy below. **No strong odor when leaves are bruised.**

Bark: Brown to dark gray, deeply fissured or sometimes broken into thick scales.

Twigs/Buds: Twigs very stout, brown, hairless. **Leaf scars round and raised above twig surface.** Terminal bud absent; side buds small (to 1/4 inch), round, several scales.

Flowers/Fruit/Seeds: Flowers bisexual, large (to 2 inches), white, showy, in spreading clusters, appear after leafing in late May to June. Fruit long (to 1 1/2 feet, 1/2 inch diam.) capsules; round, green then brown; split in halves at maturity to reveal several winged, hairy, buoyant seeds, each about 1 inch long, and wind-scattered long after leaves drop.

Habitat: Low woods originally. Widely planted in a variety of habitats.

Range: Originally in only a few counties of southwestern Indiana. Throughout state now, due to planting.

Comments: Wood light brown, very light-weight, easily worked, and has pretty grain; very durable in soil. Once used for fence posts; now by woodworkers. Catalpa sphinx caterpillar frequently eats leaves of catalpa trees; caterpillar a prized fish bait. Also called Hardy catalpa.

Section 2:
43 Introduced Species and Species Primarily of Shrub Size

Ginkgoaceae
> *Ginkgo biloba,* Ginkgo, Maidenhair tree

Pinaceae
> *Picea abies,* Norway spruce
> *P. pungens,* Colorado blue spruce
> *Pinus resinosa,* Red pine
> *P. sylvestris,* Scotch pine
> *P. echinata,* Shortleaf pine, Yellow pine
> *P. nigra,* Austrian pine

Platanaceae
> *Platanus orientalis,* Oriental planetree

Hamamelidaceae
> *Hamamelis virginiana,* Witch-hazel

Ulmaceae
> *Ulmus pumila,* Siberian elm, Dwarf elm

Moraceae
> *Morus alba,* White mulberry

Betulaceae
> *Alnus incana,* Speckled alder
> *A. glutinosa,* Black alder, European alder

Salicaceae
> *Populus alba,* White poplar
> *P. nigra* var. *italica,* Lombardy poplar
> *Salix amygdaloides,* Peach-leaf willow
> *S. alba,* White willow
> *S. fragilis,* Crack-willow
> *S. babylonica,* Weeping willow
> *S. discolor,* Pussy-willow

Mimosaceae
Albizia julibrissin, "Mimosa," Silk-tree

Elaeagnaceae
Elaeagnus umbellata, Autumn olive

Cornaceae
Cornus alternifolia, Pagoda dogwood

Celastraceae
Euonymus atropurpureus, Wahoo

Rhamnaceae
Rhamnus caroliniana, Carolina buckthorn
R. cathartica, Common buckthorn

Sapindaceae
Koelreuteria paniculata, Golden-rain tree

Hippocastanaceae
Aesculus hippocastanum, Horse-chestnut

Aceraceae
Acer platanoides, Norway-maple

Anacardiaceae
Rhus glabra, Smooth sumac
R. copallinum, Shining sumac
Toxicodendron vernix, Poison-sumac, Swamp sumac

Simaroubaceae
Ailanthus altissima, Tree of heaven, Ailanthus tree

Rutaceae
Zanthoxylum americanum, Common prickly ash
Ptelea trifoliata, Common hop-tree

Araliaceae
Aralia spinosa, Hercules' club

Oleaceae
Forestiera acuminata, Swamp-privet

Bignoniaceae
Paulownia tomentosa, Empress-tree, Royal paulownia
Catalpa bignonioides, Southern catalpa

Caprifoliaceae
Viburnum lentago, Nannyberry
V. rufidulum, Southern black haw
V. prunifolium, Black haw

Rosaceae
Pyrus calleryana, Callery pear

GINKGO; MAIDENHAIR TREE

Ginkgo biloba L.

Distinguishing Features: A handsome medium-sized tree (to 80 feet tall, 2+ feet diam.) with an ascending crown of spreading branches. Its **fan-shaped leaves (2–3 1/2 inches wide) are borne on medium, flattened leafstalks. Leaf blades usually deeply notched or cleft in center of outer margin** (hence *biloba*). **Veins of leaves repeatedly divide** (dichotomously), a characteristic shared by no other Indiana tree. Leaves turn a bright yellow in fall, then drop like drifts of gold

coins. **Stout twigs have spur shoots** (shortened, thickened stems), which bear squat broad buds that contain immature leaves and flowers. Drooping flowers are unisexual on separate trees. **Female trees bear plum-like fruits** (to 1 inch long); turn yellow-orange when ripe; the thin, sticky, pulpy flesh, which surrounds a single large white seed, has a very foul odor.

Comments: It is the only surviving member of its family since geologic time. Widely planted as an ornamental, it is frequently encountered at residences and in parks. Its soft, light wood is similar to Eastern white pine, but not used commercially. Native of Asia and frequently planted there. An extract from the tree is used widely as a purported stimulant, invigorator, or sexual enhancer.

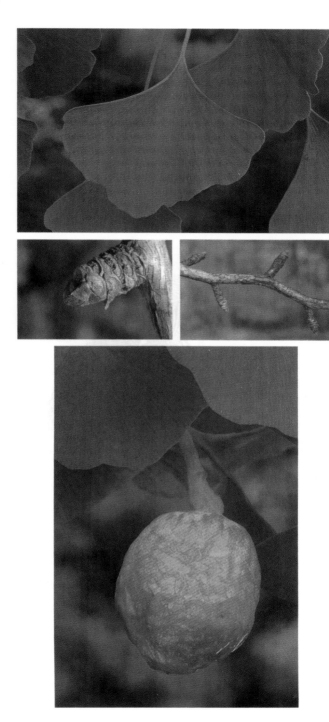

NORWAY SPRUCE

Picea abies (L.) Karsten.

Distinguishing Features: A medium to large **evergreen tree** (to 80 feet tall, 2 feet diam. here), with a triangular crown of somewhat drooping branches. **Needles** are stiff, very dark green, **squarish in cross section;** sharp-pointed; resinous and aromatic when crushed; **borne singly** on a short woody stalk, usually point toward branch tip. Branchlets appear somewhat weeping; new twigs a pale tan. Buds to 1/4 inch long, pointed, soft brown. **Cones** are warm brown, **4–7 inches long, curved, with thin, papery scales.** Bark reddish-brown, scaly.

Comments: Introduced from Europe; a handsome tree that is (was) widely planted in the United States, but only rarely reproduces here. Often found in lawns of old homesteads.

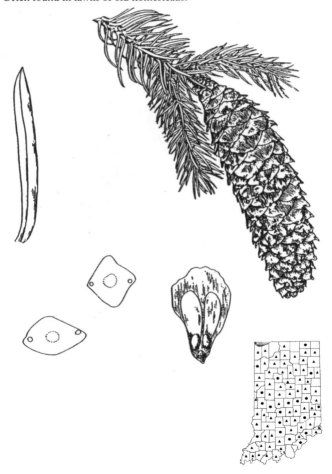

COLORADO BLUE SPRUCE

Picea pungens Engelm.

Distinguishing Features: A small to medium (here) **evergreen tree** (to 40 feet tall, 1 foot diam.) with a narrow triangular crown of short, stiff branches. **Needles (to 1 1/2 inches)** range from **a bright green to a frosted blue-green** on different individual trees; borne singly on short woody stalks; **extremely sharp-pointed; square in cross section;** pleasant, pungent odor when bruised (hence Latin name), resinous. Buds small (to 1/4 inch), pointed, green-brown, new twigs a soft brown. **Cones** 2–3 inches long; **scales papery** thin, edges scalloped to ragged. Bark dark gray, scaly.

Comments: Native to central Rocky Mountains, especially Colorado. A beautiful and much loved ornamental that is very widely planted near Indiana homes, and sometimes in parks. Also called Blue spruce.

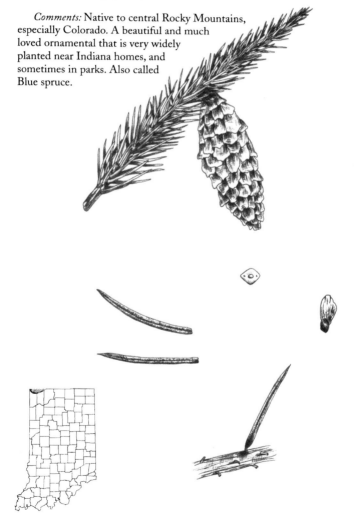

RED PINE

Pinus resinosa Aiton.

Distinguishing Features: A large **evergreen tree** (to 100 feet tall, 3 feet diam.) in its native range; usually much smaller when planted in Indiana. **Needles are 2 to a bundle, 5–6 inches long, slender, soft, flexible, dark green, lustrous; snap easily when folded in half.** Bark of mature trees with narrow furrows and broad ridges, scales thin, loose, red-brown (hence common name). A handsome tree with a symmetrical, oval crown. **Cones** oval, symmetrical, on very short stalks; scales thin, flexible, chestnut brown, **without prickles.** Seeds small (1/8 inch) with long wings (to 3/4 inch), wind-scattered. Bark a warm red-brown. Wood pale red heartwood (common name, also), medium softness and weight. Very valuable timber tree.

Comments: Native to the Great Lakes states, New England, and southeastern Canada. Often planted in Indiana, as an ornamental, and frequently encountered here. Also planted widely from 1930s to 1960s along with Shortleaf and Eastern white pine, to reforest eroded farm fields. Sometimes called Norway pine by foresters and lumbermen.

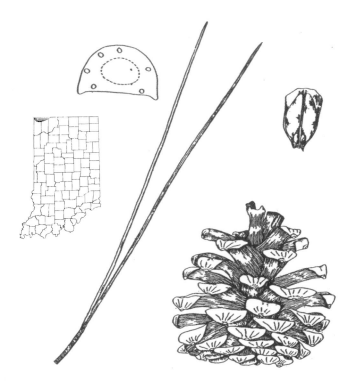

SCOTCH PINE

Pinus sylvestris L.

Distinguishing Features: A small to medium-sized **evergreen tree** (to 50 feet tall, 1+ feet diam.), **often with crooked trunk and irregular spreading crown. Bark scaly, bright orange-red;** becomes darker on older trees. **Needles** are stiff, often twisted, **yellow green in bundles of 2; 1 1/2–3 inches long.** Cones usually asymmetric at base, 2–4 inches long, flat scales with a single prickle.

Comments: Wood medium heavy; sparingly available and little used in the United States. Scotch pine is one of the most common species grown for Christmas trees in Indiana. Native of Scotland and Europe, where it is an important timber species; uncommonly regenerates here naturally. Also called Scot's pine.

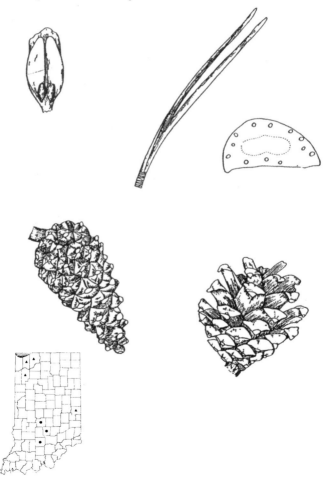

SHORTLEAF PINE; YELLOW PINE

Pinus echinata Miller.

Distinguishing Features: A large **evergreen tree** (to 80 feet tall, 2+ feet diam.) **with usually 2, sometimes 3, needles per bundle, the individual leaves 3–5 inches long.** The bark of large trees is broken into large, irregular dark brown scaly plates with cinnamon-red scales. **Cones woody, 1 1/2–2 1/2 inches long, the flat scales tipped with a prickle.** Seeds brown mottled with black (3/16 inch long), single-winged, wind-scattered. Wood is hard, strong, and heavy for a conifer. A valuable timber tree in the southern United States.

Comments: Apparently the tree was never native to Indiana. It has been widely planted in southern Indiana for more than 80 years, and is now commonly encountered on suitable sites.

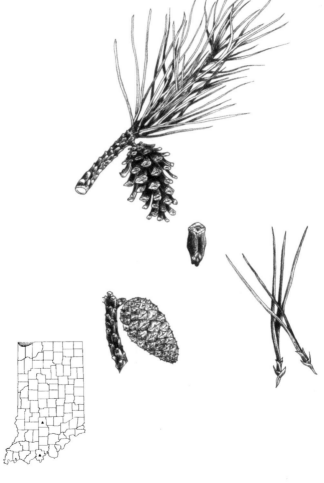

AUSTRIAN PINE

Pinus nigra J. Arnold.

Distinguishing Features: A medium-sized **evergreen tree** (to 75 feet tall, 2 feet diam. here), **with very dark green needles** (hence species name, *nigra*), and stout branches forming a pyramidal crown. **Long needles (to 6 inches) are 2 to a bundle; resemble those of Red pine (which see) except those of Austrian pine are darker, heavier, and do not break cleanly when bent double.** Woody cones are 2–4 inches long; stiff scales tipped with a tiny prickle. Bark very dark, scaly.

Comments: A European species grown fairly widely as an ornamental in North America. Not known to reproduce here.

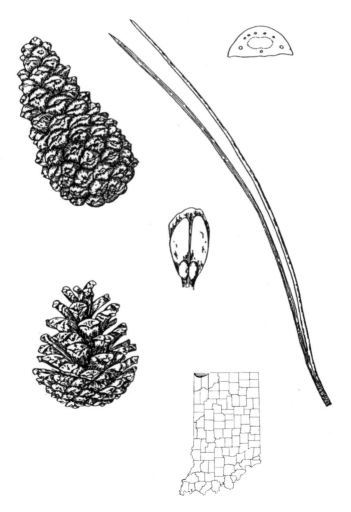

ORIENTAL PLANETREE

Platanus orientalis L.

Distinguishing Features: A medium-sized deciduous tree (to 50 feet tall; 2 feet diam. here) with broad spreading crowns and leaves alternate on twig. **Leaves lobed hand-like, but with deeper cuts than American sycamore;** borne on long (to 4 inches) leafstalks; **hollow leafstalk base covers cone-shaped bud. Globular seed heads are in strings of 2–4** on a long (to 3 inches) stalk. Seeds plumed with silky tufts are wind-scattered. **Bark mottled, greenish brown;** often peels off as large, thin scales.

Comments: Both this tree and the London planetree, *Platanus X hybrida* Brot.—a fertile, putative hybrid between the American sycamore and the Oriental planetree—are planted occasionally as street trees because they better withstand air pollution and micro-climatic extremes that damage American sycamores.

WITCH-HAZEL

Hamamelis virginiana L.

Distinguishing Features: A large shrub or small tree (to 25 feet tall, <1 foot diam.) with **a misshapen, spreading, often scraggly crown.** Of special interest because **its delicate flowers, with their 4–8 strap-like yellow petals, expand in autumn or early winter.** Lustrous black seeds are forcefully discharged in the autumn from small 2-beaked capsules, which mature from the previous year's flowers. **Leaves** alternate on short leafstalks are **oval** (to 5 inches long, 1/2 as wide) **with a lopsided base; leaf edges have widely spaced rounded teeth.** Dull green above, somewhat lighter and hairy below; turn a soft yellow in autumn. Buds stalked, pointed, to 1/3 inch. Bark thin, often scaly; exposes reddish-purple inner bark. Wood fairly heavy.

Comments: Encountered infrequently in moist woods, ravines, or along streams. Trees are usually too small to use, but forked twigs are believed, by those superstitious, to be useful to "witch" for underground water, or even gold, hence name. **Tree fragrant;** sap used for liniment, toilet water, or lotions.

SIBERIAN ELM; DWARF ELM

Ulmus pumila L.

Distinguishing Features: Medium to large tree (to 60 feet tall, 2+ feet diam.) with an open, spreading, often very twiggy, crown. **Leaves alternate on twig on very short leafstalks; small** (to 2 1/2 inches long, 2/3 as wide), **oval, unequal at base; edges singly toothed, surface slightly rough to touch.** Bark smooth, then furrowed and ridged when mature; dark gray. Flowers tiny, in clusters before leafing; **fruit greenish, winged, wafer-like with single seed;** wind-scattered.

Comments: Native to Siberia and northern China. Once widely planted near new housing for its fast growth and quick shade; less so today. Short-lived and subject to insect and disease pests. Wood medium heavy, tough, hard to split; of little value. Chinese elm (*Ulmus parvifolia* Jacq.) is a closely similar introduced species, but essentially all Indiana specimens are believed to be Siberian elm. Occasionally reproduces here, and mildly invasive.

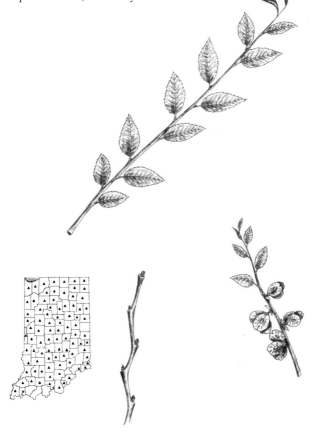

WHITE MULBERRY

Morus alba L.

Distinguishing Features: Small to medium tree (to 50 feet tall, 1+ feet diam.) with broadly rounded crown of short branches. **Leaves variable in shape; edges toothed; lower surface without hairs. Fruit multiple; a white or pinkish berry.**

Leaves: Alternate, simple, borne singly on leafstalks to 2 inches long. Blades to 4 inches long, nearly as broad; edges coarsely round-toothed and unlobed or two to several lobed; dark green above, paler below; both surfaces smooth. Turn yellow in autumn.

Bark: Thin, light brown tinged with orange; divided into long, scaly plates on mature trees.

Twigs/Buds: **Twigs slender,** yellowish, smooth to hairy; **exude milky sap when broken.** Buds small (1/6 inch long), pointed, with smooth red-brown scales.

Flowers/Fruit/Seeds: Flowers unisexual, appear as leaves unfold; sexes on same or different trees; clustered on spikes, female short, male to 2 inches, drooping. Fruit multiple "berries" (from several individual flowers); white to pink (rarely red or purple) when ripe. Edible, very sweet. Tiny hard seed in each fruit section. Bird-scattered.

Habitat: Woods edges, roadsides, fencerows, woody old fields, urban vacant lots.

Range: Throughout Indiana.

Comments: Introduced from Asia, but now widespread and reproducing naturally. Yellow-brown wood is lightweight but durable, coarse-grained. Used for fence posts. Leaves once used to feed silkworms in Asia.

SPECKLED ALDER

Alnus incana (L.) Moench.

Distinguishing Features: A tall colonial shrub or small tree (to 30 feet tall, 6 inches diam.) with broad irregular crown. Bark usually smooth, reddish-brown, gray-tinged, with grayish dots (hence name). **Leaves broadly oval 2–4 inches long, doubly toothed to minutely lobed.** Turn yellow-rusty in autumn. **Buds stalked,** dark red-brown, blunt-tipped, with 2–3 overlapping scales. **Pith of twigs triangular** in cross-section. Flowers are catkins borne separately on same tree; male long, slender (1–1 1/2 inches) often paired, female clustered on stalk, rounded (1/2 inch). **Fruits are hanging cones (1/2–3/4 inch long) in clusters,** scaled, each scale containing a tiny, nearly round, winged seed. Wind-scattered in autumn.

Comments: Primarily confined to northernmost tier of counties from Lagrange County westward. Grows in low ground on stream borders, in swamps or sloughs. Often stools out from stem bases, forming thickets. Small size renders it of no importance as lumber or fuel, but wood very hard, close-grained, and takes a high polish. Heartwood red-brown; sapwood whitish. Could be used ornamentally on suitable sites.

BLACK ALDER; EUROPEAN ALDER

Alnus glutinosa (L.) Gaertner.

Distinguishing Features: A small to medium-sized tree (to 30 feet tall, 6 inches diam.) with a narrow open crown. **Leaves alternate** (2–5 inches long, 1/2 as wide); **widest above the middle, blunt to almost notched at the tip;** edges coarsely toothed; dark green and smooth above, paler and hairy below. Flowers unisexual on same tree; male catkins drooping (to 5 inches long), female shorter (to 1/2 inch) at branch tips. **Fruits are small woody cones** (to 1 inch); seeds winged, wind-scattered in autumn.

 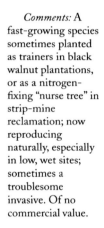

Comments: A fast-growing species sometimes planted as trainers in black walnut plantations, or as a nitrogen-fixing "nurse tree" in strip-mine reclamation; now reproducing naturally, especially in low, wet sites; sometimes a troublesome invasive. Of no commercial value.

WHITE POPLAR

Populus alba L.

Distinguishing Features: Medium tree (to 60 feet tall, 2 feet diam.) with a rounded crown of crooked branches and coarse twigs. Buds pointed (to 1/4 inch), soft, hairy, not sticky. Bark of twigs greenish-gray; **trunks often multiple, white to greenish gray or silvery gray with maturity; becomes dark gray and fissured at base.** Leaves alternate on twig on long, flattened leaf stalks (flutter in breezes); broadly oval with irregular, wavy-toothed edges to maple-like lobing; **dark green above, usually woolly white below** (hence common name). Flowers before leafing in long, slender male and female catkins; fruits are small (to 1/4 inch) capsules, also on drooping catkins. Wind-scattered.

Comments: Introduced from Europe and Asia. Quite widely planted as an ornamental; now frequently reproduces naturally, often by root suckers surrounding the parent tree. Invasive into prairies and old fields. Wood light, soft, weak, difficult to split; of no commercial value. Also called Silver poplar.

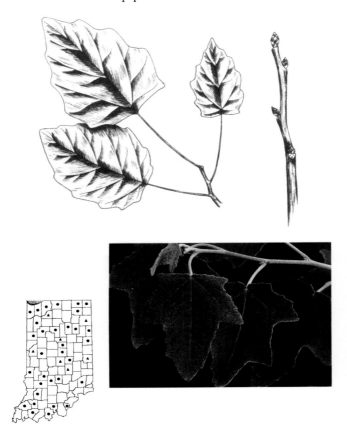

LOMBARDY POPLAR

Populus nigra var. *italica* Du Roi.

Distinguishing Features: A distinctively shaped, **medium-sized tree (to 50 feet tall, 1+ feet diam.), with a slender, columnar crown of short, sharply ascending branches.** Easily recognized from a distance by its silhouette. Leaves alternate on twig with long (to 3 inches), flattened leafstalks. Rustles with breezes. Blades triangular to diamond-shaped (2–3 inches long, almost as wide), with rounded marginal teeth. Flower and fruits (capsules) in long (to 3 inches), drooping catkins. Bark greenish-gray on younger stems, then black and furrowed on old trunks.

Comments: Introduced from Europe, possibly as clones from a single genetically unique individual found near Lombardy, Italy (hence common name). Once widely planted in the United States as an ornamental, property screen, and windbreak; less used today. Apparently seldom reproducing here spontaneously. Scale insects, borers, wind breakage, and so forth, soon take a toll on plantings.

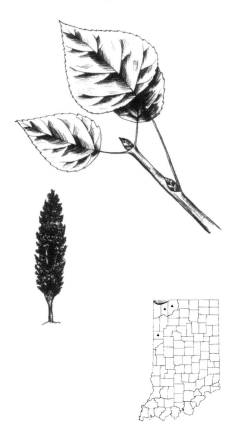

PEACH-LEAF WILLOW

Salix amygdaloides Andersson.

Distinguishing Features: A small tree (to 40 feet tall, 1 foot diam.), with a yellow-green spreading crown of drooping branchlets; somewhat weeping. Bark of old trees nearly black, ridged, and furrowed. **Leaves peach-like, alternate on twig on short stalks; long** (2–5 inches), **narrow, long pointed;** finely and closely toothed edges; pale green above, grayish below; mid-vein yellow to orange. Flowers in long (for willows) catkins (to 2 inches); fruit a cluster of capsules.

Comments: Found along stream courses and wet sites; primarily in northern half of the state. Wood soft, weak; of no commercial value.

WHITE WILLOW

Salix alba L.

Distinguishing Features: A small tree (usually to 40 feet tall, 1+ feet diam.; on occasion much larger), with a short trunk and a spreading crown of olive-green branches, and furrowed, dull brown bark. **Leaves** long (to 4 inches), narrow, and long-pointed; **alternate on very short leafstalks; finely toothed edges; dark green above, white-hairy below.** Flowers in short (to 1 inch) catkins with leaves. Fruit a cluster of capsules; seeds wind-scattered.

Comments: A native of Europe, widely planted by the settlers for quick shade and erosion control at their homesteads; now reproduces naturally. Found at scattered sites across the state.

CRACK-WILLOW

Salix fragilis L.

Distinguishing Features: A small to medium-sized tree (to 40 feet tall, 1+ feet diam.), with a broad crown of primarily upright branches; its **brittle green to red-brown twigs readily break off** (hence common name). Large trees have heavily ridged dark brown to black bark. **Leaves, alternate on short stalks; narrow, long** (to 5 inches, 1/4 as wide); smooth, dark green; edges finely toothed. Flowers in long (to 3 inches) catkins; fruits are clusters of capsules; seeds plumed and wind-scattered.

Comments: Native to Europe; introduced early by settlers for ornament, quick shade, and erosion control along streams. Scattered in Indiana, apparently largely in northern third of the state. Also called Brittle willow.

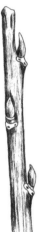

WEEPING WILLOW

Salix babylonica L.

Distinguishing Features: A medium tree (to 50 feet tall, 2+ feet diam.) with **long slender, whip-like, olive to yellow to pale orange branchlets, usually drooping gracefully in a weeping habit** (hence common name). Gray bark becomes thickened and fissured on old trees; smooth, yellowish on seedlings. Leaves simple, alternate on twig, leafstalks very short; blades long and narrow, taper to a long point, edges finely saw-toothed. Buds small (<1/4 inch), narrow-pointed, flattened with a single scale; terminal bud absent. Flowers with leafing; borne on catkins to 2 inches long. Fruits are small, dry capsules; contain many tiny seeds each with long, silky, white hairs. Wind-scattered.

Comments: Introduced from Asia and eastern Europe as an ornamental. Escapes occasionally by brittle twigs being scattered, then rooting at wet sites. A moisture-loving tree that propagates easily from cuttings. Wood light, soft, weak, nearly white; of no commercial value.

PUSSY-WILLOW

Salix discolor Muhl.

Distinguishing Features: A shrub that sometimes becomes a small tree on favorable sites (to 20 feet tall, 6 inches diam.), with an open rounded crown of stout branches (for a willow). **Leaves long and narrow to elliptical** (to 5 inches long, 1/3 as wide), short-pointed; light green above, silvery-hairy below; edges with wide-spaced teeth; mid-vein yellow. **Flowers late winter to early spring; the male flowers in dense, soft, and silky catkins** (hence common name).

Comments: A much-loved ornamental; often the first flowering woody species to break winter's drear. Grows in wet sites around bogs, lakes, stream courses, and swamps. Widespread throughout most of Indiana in suitable habitats.

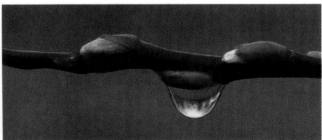

"MIMOSA"; SILK-TREE

Albizia julibrissin Durazz.

Distinguishing Features: **Small trees** (to 50 feet tall, 1 foot diam.), frequently with multiple trunks; handsome ornamentals **with large twice-compound, feather-like (bipinnate) leaves,** and a wide-spreading graceful crown. Leaves consist of a central axis, 5–15 side branches, each with 15–30 pairs of small, narrow, inequilateral leaflets with smooth edges and pointed tips. **Flowers clustered into fluffy pink (silk-like) heads (hence name);** the central flower much larger than the others. **Fruits are thin, flat legumes (pods),** 3–8 inches long and 3/4–1 1/2 inches wide, each containing several hard, flattened seeds. Thorns not present on tree.

Comments: An attractive tree that is native to subtropical Asia from Iran to Japan. Widely planted ornamentally in southern United States and now widely invading into yards, parks, abandoned fields, and along roadsides. Becoming quite common in Indiana in counties south of U.S. highway 50.

AUTUMN OLIVE

Elaeagnus umbellata Thunb.

Distinguishing Features: A shrubby, **typically much-branched small** tree (to 25 feet tall, 4–6 inches diam.); often leaning, twisted, or distorted, **with many thorny branches which form a compact crown. Leaves are** lanceolate or oblong, 2–3 inches long, edges smooth; **dark green above and covered with silvery-white scales below, giving an attractive sheen in brisk winds.** Flowers 4-parted, but petals absent. **Fruits red, dotted with pale scales,** juicy and edible to wildlife, which scatter them widely.

Comments: With myriad edible fruits and formidable thorny cover, this tree is highly attractive to birds for food, nesting cover, and shelter. This exotic is native to southern Europe and central Asia. It has been introduced into the Midwest as a highway border planting, and early on as a wildlife food and cover species. Today it is widespread, highly invasive, and almost impossible to eradicate without drastic measures.

PAGODA-DOGWOOD

Cornus alternifolia L. f.

Distinguishing Features: Small tree (to 20 feet tall, 4 inches diam.), or sometimes an erect shrub. **The only dogwood in Indiana with alternate leaf arrangement;** simple, oval leaves with wavy or smooth edges often clustered near twig tips; leaves with pointed tips (3–4 1/2 inches long, 2–3 1/2 inches wide), leafstalk to 1 1/2 inches. **Flowers several, small,** crowded into round-topped clusters, **white with 4 narrow petals. Fruit dark blue spherical berries (to 1/3 inch), borne on red stalks;** single hard pit. Leaf buds small, acute; flower buds spherical or vertically flattened.

Comments: Pagoda-like shape (hence name) is formed by horizontal branches in progressively smaller tiers. Grows in rich, moist woods. Wood is heavy, hard, close-grained, heartwood brown. Too small for commercial use except for tool handles, novelties. Is now gaining favor as an ornamental. Also called Alternate-leafed dogwood.

WAHOO

Euonymus atropurpureus Jacq.

Distinguishing Features: An erect shrub or small tree (to 20 feet tall, 4 inches diam.), which forms colonies by vigorous root sprouting. **Branches green, 4-angled with longitudinal ridges.** At maturity, the bark of the trunk becomes an attractive dark gray with deep-red fissures. **Leaves simple, opposite,** elliptical to oblong, edges finely toothed, tips long-pointed; leafstalks 1/4–3/4 inch long. **Flowers** small, perfect, 4-parted, **brownish-purple,** in branched clusters. **Crimson to reddish-purple fruits are profuse, 4-parted, open "bittersweet-like" in autumn;** occur densely in bright fiery clusters, hence another common name, Burning bush. Nutlets 1–2 in a cell, about 1/4 inch, light brown. Fruits often persist on tree until winter, or until consumed by wildlife.

Comments: A fast-growing tree with attractive flowers and fruits that make it a desirable ornamental. Grows well in moist woods under moderate to dense shade. Bark of root formerly used in medicines. Wood nearly white, dense, and hard. Also called "hearts-a-bustin'-with-love" by the pioneers for the exposed, bright red opened fruits during autumn.

CAROLINA BUCKTHORN

Rhamnus caroliniana Walter.

Distinguishing Features: A tall, erect shrub or small tree (to 30 feet tall, 6 inches diam.), which **grows on lower slopes to ridges in a few far-southern** Indiana counties near the Ohio River. **Leaves alternate, simple, shiny, elliptical** (to 5 inches long, 2 inches wide); on short (to 1 inch) leafstalks, tips blunt, bases rounded; edges smooth to slightly wavy-toothed. Flowers after leafing, greenish-yellow, inconspicuous, in flat-topped clusters. **Fruits round** (to 1/3 inch diam.); **shiny-red, then black later;** 3 nutlets.

Comments: It is a nice ornamental with its shiny green foliage and red fruit. Leaves turn orange-brown and hold on trees until well into the winter. Fruits attractive to a number of bird species. Wood brittle, of no value.

COMMON BUCKTHORN

Rhamnus cathartica L.

Distinguishing Features: A small tree (to 25 feet tall, 8 inches diam.) with **some branches ending in short thorns. Leaves alternate to opposite, broadly elliptical** (to 2 1/2 inches long, 1 1/2 inches wide); short-pointed, base rounded to tapered; bright green, **veins prominent; edges toothed;** persist on branches into winter. Flowers in clusters in leaf axils after leafing; on separate trees or mixed. **Fruit round** (to 1/4 inch), **fleshy, black, bitter;** has 3–4 seeds.

Comments: Native of Europe and Asia; widely introduced and aggressively invasive here in a variety of habitats. **It should not be planted.** Fruit has been used medicinally (e.g., as a purgative); sap can be used as a dye.

GOLDEN-RAIN TREE

Koelreuteria paniculata Laxm.

Distinguishing Features: A small tree (to 25 feet tall, <1 foot diam.), often with clumped double or triple trunks. Compact limbs and dense foliage form a full round-topped crown. **Compound leaves have 7–11 broad leaflets** on short leafstalks. Leaflet blades are oval, smooth, rich dark green above, yellow-green below; edges coarsely toothed to notched. In June, mature trees are covered with **showy sprays of golden flowers,** which later sprinkle the ground with individual flowers, hence the golden-rain tree. **Fruits are hard nutlets, each covered by** a pyramid-shaped inflated **papery bladder** (1 1/2–2 inches long); at first pale green, then brown. The clusters of fruits remain attached until late autumn; largely wind-scattered.

Comments: A handsome ornamental in all seasons. Originally called the "gate tree" from being planted at the gate of William McClure's residence in New Harmony, where he introduced the species to Indiana. Native of southeast Asia; occasionally escapes here. Wood hard, heavy; trees too small to be of value.

HORSE-CHESTNUT

Aesculus hippocastanum L.

Distinguishing Features: Sometimes grows to rather large size (to 60–70 feet tall; 2 feet diam.) with a broad, rounded crown. In May it has **large clusters of showy white to cream flowers, mottled with red.** The smooth bark becomes gray and scaly with age. **Leaves, with long, heavy leafstalks, are opposite on the twigs, hand-like compound, with seven (usually) large obovate leaflets.** Fruit large (2–3 inches diam.) spiny capsules with a thin leathery husk. The **"buckeye" seed,** 1–1 1/2 inches across, is a rich glossy brown with large white "eye." Squirrel-scattered.

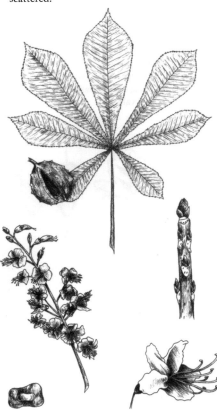

Comments: Native of southern Asia; widely planted in eastern United States as a handsome ornamental. Seedlings sometimes grow under planted trees; uncommonly establishes in natural habitats. Wood and uses similar to that of Ohio-buckeye. This, apparently, is the "spreading chestnut tree" the village blacksmith stood under in Longfellow's poem.

NORWAY-MAPLE

Acer platanoides L.

Distinguishing Features: A medium-sized tree (to 60 feet tall, 2 feet diam.) with a full, rounded crown. **Leaves resemble rather closely those of sugar maple, but leafstalks exude a milky, sticky sap when bruised** and squeezed. **Lovely clusters of greenish-yellow flowers are borne on long stems** after the leaves. **Paired, winged seeds are widely spread** (1 1/2–2 inches across), appear in summer, and wind-scattered. The thinly furrowed bark is nearly black. Wood heavy, brittle, of little use, except for firewood.

Comments: An introduced species from Europe and Asia; now reproducing naturally here, and becoming a troublesome invasive. Frequently planted ornamentally in tree rows, lawns, and parks of cities. As a street tree, its dense, rounded crown produces excellent shade. However, its invasive abilities render the species undesirable as an ornamental.

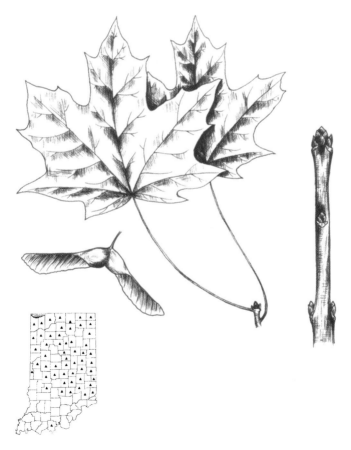

SMOOTH SUMAC

Rhus glabra L.

Distinguishing Features: Primarily a vigorously **colonial shrub to small tree** (to 20 feet tall, 6 inches diam.) **that often grows in rounded thickets** along roadsides, woodland margins, on abandoned farmlands, or in moist bottomlands. **Twigs very smooth,** hairless, slightly shiny, **somewhat triangular. Leaves alternate, featherlike, compound, toothed, leaflets usually 11–31,** sharp-pointed, lance-like; dark green above, whitish below. **Turn bright red in autumn.** Flowers greenish-yellow, very numerous in large terminal clusters. Male and female usually on separate plants. **Fruits are clusters of deep red** berries (to 1/8 inch diam.); sour to taste; contain a single brown seed.

Comments: An important successional species on abandoned lands of the Midwest. Fruit can be steeped to make "Rhus-ade," and an extract for flavoring "lemon-like" pies. Wood light yellow-brown, beautiful grain for turning. Should be used more ornamentally, as it has an attractive form, beautiful foliage, and attractive fruits.

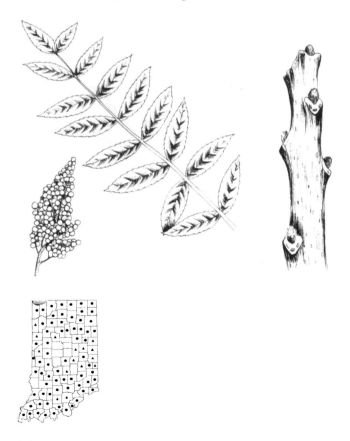

SHINING SUMAC

Rhus copallinum L.

Distinguishing Features: Slender shrub or small tree (to 20 feet tall, 4–5 inches diam.) with a short trunk and a rounded crown of alternate compound leaves. Leaves ladder-like with 9–21 leaflets, ovate-lanceolate, sharp-pointed at tips, all **leaflets attached to a winged leafstalk.** Shiny green and smooth above, paler and hairy beneath. **Leaves turn a deep red or wine color in autumn.** Flowers many in clusters, tiny, greenish-yellow; male and female on separate plants. **Fruits in fall are clusters of red berries (to 1/8 inch diam.), each finely hairy,** contains a single orange oily seed. Wood is soft, coarse-grained, pale yellow-brown.

Comments: Grows on dry hills, woods borders, and in abandoned fields. **Readily distinguished by winged leafstalks.** Also called Winged sumac.

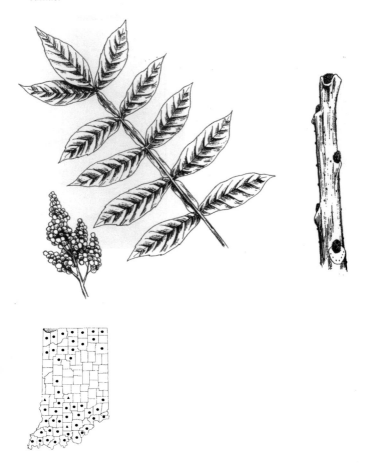

POISON-SUMAC; SWAMP SUMAC

Toxicodendron vernix (L.) Kuntze.

Distinguishing Features: **The most poisonous plant to the touch of any in Indiana;** Toxicodendron is literally "poison tree." A tall, erect shrub to small tree (to 20 feet tall, 6 inches diam.) with an attractive branching crown and smooth, gray bark which resembles that of beech trees. **Leaves large, ladder-like, compound with 7–13 dark green, shiny,** pointed, toothless **leaflets. Leaves turn a beautiful scarlet in autumn.** Flowers in clusters, small greenish-yellow, male and female on the same or separate plants. **Fruits white,** round (to 1/2 inch diam.) **in grape-like clusters** in early fall. Eaten greedily by cardinals in winter.

Comments: Typically grows in oxygen- and nutrient-deficient soils of bogs and fens, most commonly in northern Indiana. Contains a potent irritant, urushiol (a phenol), which causes rashes or blisters in many people, even from brushing against a single leaf. **Wash exposed skin soon and thoroughly with liquid detergent if in contact with the plant.**

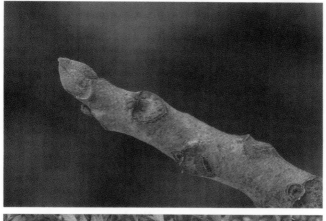

TREE OF HEAVEN; AILANTHUS TREE

Ailanthus altissima (Mill.) Swingle.

Distinguishing Features: A medium to large tree (to 80 feet tall, 2+ feet diam.), with a loose, open crown. Often grows in clumps. **Leaves alternate, feathery compound with many leaflets; very stout twigs. Branches have large leaf scars and large pith. Flowers and winged fruits in dense clusters. Strongly malodorous,** especially male trees. Frequently found in waste places.

Leaves: Alternate on stout twigs; very large, compound with many (11–41) leaflets; single terminal leaflet; leaflets long (2–6 inches, 1/3 as wide), edges smooth except for glandular teeth at base; leafstalks swollen at base.

Bark: Bark thin, very dark gray, somewhat roughened on old trees.

Twigs/Buds: Twigs very stout with large pith; easily broken; yellowish to red-brown; feel smooth, to soft-downy. Leaf scars large, heart-shaped; many "pores" on bark of twigs. End buds absent; side buds small, half-round, brownish, fuzzy.

Flowers/Fruit/Seeds: Flowers unisexual on separate trees, small, yellow-green, many in large clusters. Male flowers have foul odor. **Fruits dry, papery; the single seed in the center of an oblong, twisted wing about 1 1/2 inches long. Hang in large, dense clusters until late winter.** Wind-scattered.

Habitat: Thrives in waste places, especially in blighted areas of cities. Does especially well on fertile, moist soils, and rocky limestone ledges along the Ohio River. Now quite invasive.

Range: Native of northern China. Introduced into United States as an urban ornamental in 1874. Now escaped and naturalized throughout Indiana.

Comments: An exceedingly fast-growing tree—to 10 feet in height in one year. Wood weak, vulnerable to breakage in storms; tree is of essentially no value commercially, and only rarely to wildlife, but used occasionally as an ornamental. Also called Stink tree for the disagreeable odor of crushed leaves or male flowers.

COMMON PRICKLY ASH

Zanthoxylum americanum Miller.

Distinguishing Features: A prickly shrub or low tree (to 15–20 feet tall, 3–4 inches diam.) with **branches armed with straight spines at the base of each leaf.** Spines are flattened at the base, persistent, and reach 1/4–1/2 inch long. Leaves alternate, compound with **5–11 ladder-like, ovate leaflets; edges smooth or nearly so;** hairy below. Flowers small, yellow-green; occur before leaves; sexes separate. **Fruit in clusters of reddish capsules;** seeds black and shiny; fruit especially, and the **entire plant, aromatic** with lemon-like odor.

Comments: Occurs in moist woods in northern Indiana, and on rocky, wooded slopes in southern Indiana. Where it is found it sometimes root sprouts into dense thickets. Once much used in medicines; formerly called "toothache tree," apparently from its use as a presumed pain reliever.

COMMON HOP-TREE

Ptelea trifoliata L.

Distinguishing Features: An erect shrub or small tree (to 20–25 feet tall, 4–5 inches diam.) with rounded crown. **Leaves alternate, divided into three sessile leaflets,** each ovate, tapered-pointed at tip, edges smooth or nearly so; leafstalks long (2 1/2–4 inches). Flowers small, greenish-white in large clusters. **Fruit a distinctive round (to 1 inch across), winged-veined samara** (hence Wafer-ash, another common name); pale green early, later a brown paper bag color; remains on tree until late winter. **Twigs dark brown, often "warty"; leaf scars horse-shoe-shaped.**

Comments: Widespread across Indiana; often found along stream courses, on rocky wooded slopes, along roadsides, in fencerows, and crests of Indiana dunes. Wood hard, heavy, close-grained, yellow-brown; trees too small to have value. Could be used ornamentally. Juice once used to make a foul-tasting tonic for use as a quinine substitute.

HERCULES' CLUB

Aralia spinosa L.

Distinguishing Features: A tall shrub or small tree (20–30 feet tall, to 6 inches diam.). **Stems stalky,** with few stout branches and a large central pith; armed **with many short (to 3/8 inch), stout spines, usually at nodes. Leaves alternate, doubly compound, huge** (to 4 feet long and 3 feet wide), leafstalks often with prickles, and have large, clasping bases; leaf edges sharp-toothed to smooth. Buds large (to 3/4 inch), conical, blunt; leaf scars nearly encircle twig. Flowers tiny, occur in August, cream-white, in large spreading clusters. **Fruit juicy, purple-black berries** (1/2 inch), mature in September/October; eaten readily by birds and mammals.

Comments: Species often root sprouts into extensive colonies, most commonly on fertile soils along streams, occasionally on dry uplands. With umbrella-shaped crown, it could be a promising ornamental. Formerly used in medicine, and to color hair black. Too small and pithy to have any value for wood. Also called Devil's-walkingstick.

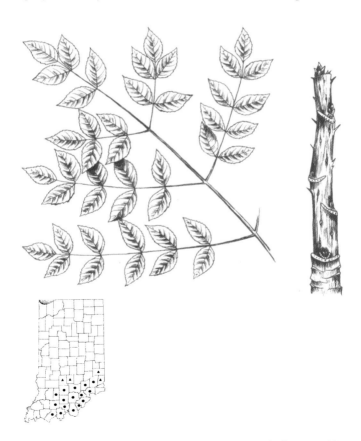

SWAMP-PRIVET

Forestiera acuminata (Michx.) Poiret.

Distinguishing Features: A small tree (to 30 feet tall, 5 inches diam.) with opposite, simple leaves, and an irregular, spreading crown. Found naturally in Indiana only in low wet woods of far southwestern Indiana, usually in bayous and swamps. **Leaves opposite** and leafstalks about 1/2 inch long, **ovate to lance-shaped with a long-tapered tip, edges smooth near base to slightly round-toothed nearer tip.** Flowers yellowish, small, in clusters, before leafing, male and female on separate trees. **Fruits oblong, dark purple drupes** to 1 inch long; one grooved stone. Eaten by waterfowl.

Comments: Stems often deformed by yearly flood inundations of the low-ground habitat. Wood is soft, light, and weak; stems too small to be used. Local people call it Pond brush. Not an unattractive plant; can be used ornamentally on wet sites.

EMPRESS-TREE (ROYAL PAULOWNIA)

Paulownia tomentosa (Thunb.) Steudel.

Distinguishing Features: A medium-sized tree (to 40 feet tall, 1 1/2 feet diam.) that **resembles the catalpas, but can be distinguished by its densely hairy buds** (hence species name), **purple flowers, and oval, pointed capsules** (1–1 1/2 inches long). **Leaves are velvety, heart-shaped, very large** (sometimes >1 foot across), **with smooth edges;** attached oppositely to stout, warty twigs on long leafstalks (3–6 inches). Branchlets have hollow pith and small buds above large round leaf scars. Blue flowers occur before leafing in beautiful pyramids in favorable seasons; wonderfully fragrant. Pointed fruits hang castanet-like on the trees, rattling in breezes until late fall when the small, filmy-winged seeds are wind-scattered.

Comments: A handsome ornamental that is usually winter hardy only in the southern half of Indiana, where it occurs infrequently, and sometimes escapes. Wood soft, purple-brown, satiny—esteemed in Asia where the tree is native; too rare to be used here. Also called Princess tree.

SOUTHERN CATALPA

Catalpa bignonioides Walter.

Distinguishing Features: Medium-sized trees (to 45 feet tall, about 2 feet diam.) with a broadly rounded crown. **Leaves whorled, large, heart-shaped. Bark thin, scaly. Flowers large, white, showy in large clusters; fruit a long, slender, green, round "pod."**

Leaves: Opposite (in whorls of 3) borne singly on stout, long (4–6 inches) leafstalks. **Blades to 10 inches long by 6 inches wide, heart-shaped, short-pointed at tip,** edges smooth; yellow-green above, paler and softly hairy below. **Leaves have a disagreeable odor when crushed.**

Bark: Light brown, thin, separating into thin scales on the trunk.

Twigs/Buds: Twigs stout, green to purplish and hairy when young; becoming orange-brown, smooth, and with many pores with age. Leaf scars in whorls of three, elevated, round to oval; 12 or more bundle traces in ring. End bud absent; side buds small (<1/4 inch), rounded, with red-brown scales.

Flowers/Fruit/Seeds: Flowers bisexual, large, showy-white, in branched, pyramid-shaped clusters, each with yellow and purple spots; the lower corolla lobe entire. **Fruit long (to 15 inches, 1/4 inch diam.) capsules;** split at maturity to release **flattened winged seeds, rounded at end with a fringe of hairs.**

Habitat: On rich soils and stream banks in native range. Occurs now at farmsteads, along railroads, fencerows, in towns—almost everywhere.

Range: Originally in the southern United States from Florida to Texas. Now widely planted and escaped throughout Indiana.

Comments: Wood similar to that of *Catalpa speciosa* (which see). Some manuals (e.g., Swink and Wilhelm) suggest that both Catalpas be considered as one species due to difficulty in separating them consistently. Widely reproducing naturally. Leaf odor makes it less desirable as an ornamental, but foliage and flowers are showy.

NANNYBERRY

Viburnum lentago L.

Distinguishing Features: A small tree (to 25 feet tall, 5 inches diam.) with a full, rounded crown; **most commonly occurs in boggy or wet places** surrounding lakes, ponds, and swamps, primarily of northern Indiana. **Leaves opposite, simple, ovate, sharp-pointed at the tip;** edges sharply and finely toothed; **leafstalks** short (to 1 inch), **typically winged.** Buds brown-red, long-pointed, nearly smooth (to 3/4 inch long). Flowers many, small, creamy white in broad, round-topped clusters. **Fruits** (to 2/3 inch long) ripen September/October, **edible, sweet, blue-black with a whitish bloom;** contain a single smooth stone. Wood hard, strong, orange-brown; trees too small to be of value.

Comments: This small tree, attractive in all seasons, would be a desirable ornamental. Tasty fruits eaten by many song and game birds, and by humans when encountered. Also called Sheepberry.

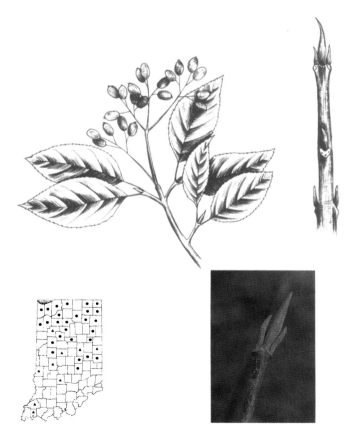

SOUTHERN BLACK HAW

Viburnum rufidulum Raf.

Distinguishing Features: An erect shrub or small tree (to 30 feet tall, 8 inches diam.), with trunk branching close to the ground, and a spreading irregular crown. It is **primarily restricted to southeastern counties in Indiana,** where it is found **on rocky, wooded slopes,** typically associated with dry-site oaks. **Leaves opposite, simple,** blades elliptic to obovate; short-pointed to rounded tips; edges sharply and finely toothed; **lower surfaces and short (to 3/4 inch) leafstalks are covered with rusty hairs.** Flowers white, small, numerous in flat-topped clusters. Fruits are blue-black, ovoid (to 2/3 inch), sweet-edible, with a single stone. **Buds are rusty-hairy,** pointed (to 1/2 inch), with two scales.

Comments: Wood heavy, hard, strong but brittle, orange-brown. An attractive small tree, sometimes planted ornamentally, and should be more so.

BLACK HAW

Viburnum prunifolium L.

Distinguishing Features: A small tree (to 25–30 feet tall, 6 inches diam.) with a compact crown; trunk usually short, crooked, branching close to the ground; bark reddish-brown, furrowed, broken into short, irregular plates. **Leaves opposite, simple, oval, tips short-pointed to blunt;** finely and sharply toothed edges; **leafstalks short** (to 2/3 inch), **slender, not winged.** Leaves similar to cherry leaves, hence the specific name. Flowers white, small, numerous in flat-topped clusters. Fruits oblong (to 2/3 inch), blue-black with whitish bloom, edible, contain a single stone. Wood heavy, hard, strong but brittle, red-brown; trees too small to be of value.

Comments: **Found throughout Indiana on hillsides, rocky slopes, woods margins, fencerows,** and sometimes abandoned fields. A handsome tree with nice flowers, fruit, nice foliage, and bright red leaves in the fall. Should be used more ornamentally.

CALLERY PEAR; ORNAMENTAL PEAR

Pyrus calleryana Decne.

Also commonly known as 'Bradford' pear, 'Cleveland Select' pear, and by other cultivar names.

Dist. Features: Small to medium size tree (to 30 ft. tall), with very showy white blooms in very early spring and striking red, maroon, and orange foliage in late fall. Various popular landscaping cultivars are distinguished by their crown structure with 'Bradford' having a short, compact, symmetrically round "lollipop" shaped crown and 'Cleveland Select' having a more elongated, conical crown. Highly invasive, spreading from ornamental landscapes into surrounding wildlands, fencerows, and highway and utility corridors at an alarming rate!

Leaves: Alternate, simple, heart to oval shaped, 1.5 to 3 inches long, on a long leaf stem. Leaf margin wavy and slightly toothed. Glossy, dark green, thick leaf surface during the summer and turning various shades from brilliant red and orange to dark maroon in late fall. In many locations and depending on fall weather, a hard freeze will kill the leaves and turn them brown before turning fall colors.

Bark: Young bark smooth, reddish brown, with lenticels, turning gray-brown with fissures and scaly ridges with age.

Twigs/Buds: Shiny reddish-brown, medium thickness, with spur shoots resembling long spines or thorns on wild trees (landscape cultivars lack these spur shoots). Buds 1/4–1/2in. long, ovate, with hair-covered/fuzzy scales.

Flowers/Fruit/Seeds: Flowers are white, 5-petaled and 3/4–1 in. across. They produce abundant blooms in very early spring (prior to most native plant leaf out or bloom) and are very conspicuous on the landscape. Flowers give off a "fishy" odor that many consider obnoxious. Fruits are small (3/8 in. diameter), round, and hard and not edible to humans. They become edible to birds only after hard freezing softens them, after which birds will eat them in large numbers and spread their seed all over the surrounding landscape.

Habitat: Readily invades open space in vacant lots, fallow fields, fencerows, on roadsides and utility corridors, at forest edges and gaps, and in most natural areas where there is sufficient sunlight on a wide variety of soils.

Range: Native of East Asia. First introduced into the U.S. in the early 20th century by USDA researchers looking for fire-blight resistant pear species that could be used as root stock for common pear (*Pyrus communis*) production. The 'Bradford' cultivar was identified as a desirable landscape variety and was subsequently widely planted as an ornamental. Other varieties were subsequently developed and were also widely planted. Wild callery pear is rapidly spreading throughout most of the eastern U.S.

Comments: Extremely invasive and spreading very rapidly in the eastern U.S. Many states have banned its sale and transplanting. Should not be planted but rather should be removed from ornamental landscapes and replaced with noninvasive or native tree species.

SPECIES EXCLUDED, EXTIRPATED, OR SUBJECT TO TAXONOMIC REVISION

The following taxa were listed in editions of *Trees of Indiana* by Charles C. Deam, some of which he considered to be distinct species, or valid subspecies. In some cases, the taxon in question no longer occurs in Indiana, or its presence here is doubtful. For others, the taxonomic status is now considered differently by most authors. For questionable cases, the nomenclature follows Gleason and Cronquist, second edition, 1991, our primary nomenclatural source.

Betula populifolia Marshall. Gray birch. This small tree with chalky-white bark that does not separate into thin layers as does Paper birch (which see) formerly occurred as a widely disjunct population in far northwestern Indiana counties. It is now considered to be extirpated from the state, although there apparently is one extant population in St. Joseph County (Ron Rathfon, personal communication, 2003), and it may be encountered as a planted ornamental.

Carya texana Buckley. Black hickory. A small hickory typical of dry sandy ridges and cliff borders has been reported in scattered locations in the Knox County region of southwestern Indiana. Presently it is on the State Endangered Species List and its status and distribution are not well known. It characteristically has nearly black, furrowed bark, yellow scales along the leafstalks, and 4-angled nuts with thin husks which split nearly to the base at maturity.

Fraxinus biltmoreana Beadle. Biltmore ash. This large ash, which closely resembles White ash, was reported by Deam as occurring quite widely in southern and central Indiana. This more typically southern form has more deeply furrowed bark and strongly hairy twigs, and often leaves. Biltmore ash is now considered by Gleason and Cronquist to be included with White ash, *Fraxinus americana*, but possibly hybridizing with Green ash, *F. pennsylvanica.*

Fraxinus lanceolata Borckhausen. Green ash, and *F. pennsylvanica* Marshall. Red ash. Formerly these taxa were considered as separate species, or in some cases, the glabrous forms of Green ash were segregated as var. *subintegerrima* (Vahl.) Fern. Gleason and Cronquist do not regard these differences as taxonomically significant, and place these closely related forms into *F. pennsylvanica* as Green ash. We are following Gleason and Cronquist's interpretation, herein.

Ilex decidua Walter. Possum-haw. Deam included this lowground species of southwestern counties in his *Shrubs of Indiana*, although, on occasion, it reaches tree size. We elected to consider it typically to be a shrub, hence it is excluded. It is characterized by alternate, remotely toothed leaves clustered at the ends of spur-like shoots, and by its red berries (Mohlenbrock, 1973).

Populus balsamifera Linnaeus. Balsam-poplar. This northern species of poplar may occur sparingly in the three counties which border Lake Michigan at low ground sites near the lake and also on the beach of the lake. Characteristics include round petioles; leaves ovate to lance-ovate, rounded bases, whitish below; catkins long-ciliate.

Quercus coccinea Muenchh. Scarlet oak, and *Q. ellipsoidalis* E. J. Hill. Northern pin-oak. Although both of these taxa are considered herein as separate species (*sensu* Gleason and Cronquist), recent taxonomic manuals question the legitimacy of their separation. Both Voss in the new *Flora of Michigan* and Swink and Wilhelm in the *Flora of the Chicago Region* consider *Q. ellipsoidalis* to be a northern expression of *Q. coccinea,* as does W. R. Overlease (personal communication, 2002). These latter interpretations may be more accurate evaluations of these two taxa, but we elected to follow the traditional separation, partly on the suggestion of Richard Jensen et al. (1984), and partly as based on observable differences between the two taxa in the field.

Quercus falcata var. *pagodafolia* Ell. Cherrybark oak was previously considered to be a variety or subspecies. Most recent manuals, including Gleason and Cronquist, have elevated this tree to full species status, *Q. pagoda* Raf. We consider it as such, herein.

Quercus X *deamii* Trel. Deam oak. This natural hybrid, purported to have occurred between *Q. macrocarpa* X *Q. muehlenbergii,* was discovered by Charles Deam in Wells County. The tree and site are now a designated state memorial. Other such individual trees should be searched for, especially in northeastern Indiana.

Sorbus decora (Sarg.) C. K. Schneider. Showy mountain-ash. Deam listed this large shrub or small tree as occurring very sparingly in the northernmost tier of Indiana counties. Apparently it has now been extirpated from the state (Homoya, personal communication, 2001). It is characterized by alternate, compound leaves (13–17 leaflets), margins finely toothed; fruit, bright red, in clusters.

Tilia heterophylla (Vent.) Loudon. White basswood. This tree, whose main range is in the mountains of the eastern United States, occurs in the counties adjacent to and near the Ohio River. Deam (1953), along with most taxonomists, considered it to be separate from *T. americana* L. American basswood. Other than the white-appearing under-leaves (from the fine, densely occurring hairs there), the two species were closely similar. Gleason and Cronquist now consider these populations to be conspecific, and we are following their lead in combining these taxa.

Ulmus parvifolia Jacq. Chinese elm. This exotic species is considered by Cope (2001) and others to be separate from *U. pumila* L. Siberian elm, also an exotic introduction. Herein, we do not consider the small-leaved introduced elms as different taxa; all such individuals, whether planted ornamentally or naturalizing, are considered as Siberian elms.

State of Indiana Record Trees by Species

Species	County	Circumference (Inches)	Height (Feet)	Point Index
Acer negundo (Boxelder)	Posey	141	62	216
A. nigrum (Black maple)	Morgan	154	82	255
A. rubrum (Red maple)	Cass	204	113	410
A. saccharinum (Silver-maple)	Monroe	263	98	388
A. saccharum (Sugar-maple)	Whitley	152	110	283
Aesculus flava (Sweet buckeye)	Jefferson	145	91	251
A. glabra (Ohio-buckeye)	St. Joseph	115	84	213
Amelanchier arborea (Downy serviceberry)	Perry	65	62	138
A. laevis (Smooth serviceberry)	Vanderburgh	32	54	92
(Co-champions)	Vanderburgh	61	24	92
Asiminia triloba (Pawpaw)	Vanderburgh	11	36	52
Betula alleghaniensis (Yellow birch)	LaPorte	102	80	198
B. nigra (River birch)	Vanderburgh	174	80	272
B. papyrifera (White birch)	Vanderburgh	100	61	176
Carpinus caroliniana (Hornbeam)	Marion	153	33	201
Carya cordiformis (Bitternut-hickory)	Perry	118	136	278
C. glabra (Pignut-hickory)	Owen	151	122	295
C. illinoensis (Pecan)	Vanderburgh	192	98	315
C. laciniosa (Shellbark-hickory)	Franklin	143	111	280

(*continued*)

State of Indiana Record Trees by Species *(continued)*

Species	County	Circumference (Inches)	Height (Feet)	Point Index
C. ovalis (Red-hickory)	Vanderburgh	82	117	214
C. ovata (Shagbark-hickory)	Wayne	150	104	273
C. pallida (Sand hickory)	—	No current champion.		
C. tomentosa (Mockernut-hickory)	Vanderburgh	84	117	214
Castanea dentata (American chestnut)	Knox	24	39	69
Catalpa speciosa (Southern catalpa)	Vanderburgh	290	85	392
Celtis laevigata (Sugarberry)	Vanderburgh	174	78	273
C. occidentalis (Hackberry)	Rush	212	105	338
Cercis canadensis (Redbud)	—	No current champion.		
Cladrastis lutea (Yellow-wood)	Monroe	98	63	169
Cornus florida (Flowering dogwood)	Vanderburgh	76	32	122
Crataegus crus-galli (Cockspur-thorn)	Tippecanoe	35	26	70
C. mollis (Downy hawthorn)	—	No current champion		
C. punctata (Dotted hawthorn)	—	No current champion		
C. viridis (Green hawthorn)	Vanderburgh	15	24	44
Diospyros virginiana (Persimmon)	Posey	110	92	217
Fagus grandifolia (American beech)	LaGrange	182	145	350
Fraxinus americana (White ash)	Scott	247	80	350

F. nigra (Black ash)	Shelby	77	97	184
F. pennsylvanica (Green ash)	Vigo	218	82	323
F. profunda (Pumpkin-ash)	Vanderburgh	112	78	204
F. quadrangulata (Blue-ash)	Washington	137	135	290
Gleditsia aquatica (Water-locust)	—	No current champion		
G. triacanthos (Honey-locust)	St. Joseph	173	105	344
Gymnocladus dioica (Kentucky coffee-tree)	St. Joseph	126	120	263
Hamamelis virginiana (Witch hazel)	—	No current champion		
Juglans cinerea (Butternut)	Fayette	217	70	308
J. nigra (Black walnut)	Fayette	182	119	327
Juniperus virginiana (Eastern red cedar)	Lawrence	120	86	214
Larix laricina (Tamarack)	—	No current champion		
Liquidambar styraciflua (Sweet gum)	Vanderburgh	154	132	309
Liriodendron tulipifera (Tulip-tree)	Daviess	271	87	379
Magnolia acuminata (Cucumber-tree)	Fayette	214	59	289
M. tripetala (Umbrella-tree)	—	No current champion		
Morus rubra (Red mulberry)	St. Joseph	216	56	284
Nyssa sylvatica (Black gum)	Pike	151	89	254
Ostrya virginiana (Hop-hornbeam)	Parke	67	93	168
Oxydendrum arboreum (Sourwood)	Perry	50	62	121
Pinus banksiana (Jack-pine)	Tippecanoe	44	49	102

(continued)

State of Indiana Record Trees by Species *(continued)*

Species	County	Circumference (Inches)	Height (Feet)	Point Index
P. strobus (White pine)	Brown	164	104	290
P. virginiana (Virginia-pine)	Martin	103	102	216
Platanus occidentalis (Sycamore)	Johnson	307	122	456
Populus deltoides (Cottonwood)	LaPorte	301	124	446
P. grandidentata (Big-toothed aspen)	Brown	85	113	212
P. heterophylla (Swamp-cottonwood)	Posey	88	78	188
P. tremuloides (Quaking aspen)	—	No current champion		
Prunus americana (American plum)	—	No current champion		
P. hortulana (Wild goose plum)	Vanderburgh	29	24	58
P. nigra (Canada-plum)	Owen	70	59	137
P. pensylvanica (Pin cherry)	—	No current champion		
P. serotina (Wild black cherry)	Owen	225	86	330
Pyrus coronaria (Sweet crabapple)	—	No current champion		
P. ioensis (Prairie crabapple)	—	No current champion		
Quercua alba (White oak)	Clay	283	83	398
Q. bicolor (Swamp white oak)	Shelby	170	155	347
Q. coccinea (Scarlet oak)	Washington	204	103	329
Q. ellipsoidalis (Northern pin-oak)	St. Joseph	90	54	161

(continued)

Q. falcata (Southern red oak)	Vanderburgh	207	76	311
Q. imbricaria (Shingle oak)	Vanderburgh	214	83	325
Q. lyrata (Overcup-oak)	Gibson	138	102	257
Q. macrocarpa (Bur-oak)	Posey	293	111	430
Q. marilandica (Black-jack oak)	Vanderburgh	80	78	170
Q. michauxii (Swamp chestnut-oak)	Jennings	285	96	411
Q. muehlenbergii (Chinkapin-oak)	Jackson	229	111	375
Q. pagoda (Cherrybark-oak)	Vanderburgh	252	95	380
Q. palustris (Pin-oak)	Greene	274	105	406
Q. prinus (Rock chestnut-oak)	Bartholomew	125	130	271
Q. rubra (Northern red oak)	LaGrange	202	115	343
(Co-champions)	Marion	243	79	343
Q. shumardii (Shumard oak)	Floyd	316	102	447
Q. stellata (Post-oak)	Vanderburgh	182	76	278
Q. velutina (Black oak)	Posey	211	92	328
Rhus typhina (Staghorn sumac)	—	No current champion		
Robinia pseudoacacia (Black locust)	Orange	170	83	260
Salix nigra (Black willow)	—	No current champion		
Sassafras albidum (Sassafras)	Harrison	172	59	245
Taxodium distichum (Bald cypress)	Knox	336	89	440
Thuja occidentalis (Northern white cedar)	Montgomery	86	42	136
Tilia americana (American basswood)	Vanderburgh	124	80	219

State of Indiana Record Trees by Species (continued)

Species	County	Circumference (Inches)	Height (Feet)	Point Index
Tsuga canadensis (Eastern hemlock)	Montgomery	101	120	233
Ulmus alata (Winged elm)	Spencer	67	104	184
U. americana (American elm)	Parke	236	112	380
U. rubra (Slippery elm)	Daviess	217	114	357
U. thomasii (Rock-elm)	—	No current champion		

Source: Division of Forestry, Indiana Department of Natural Resources, *Indiana Big Tree Registry* (2005).

Notes: Many of these record trees are not specimens that grow in natural vegetation. Instead, some are street trees in towns or cities, or occur in urban parks. Others are ornamentals at residences.

The largest individual tree currently on record in terms of Point Index is the Johnson County Sycamore at 456. The largest circumference goes to the Knox County Bald Cypress at 336 inches; the tallest recorded tree is the Shelby County Swamp White Oak at 155 feet.

Obviously these records change frequently as new or larger individuals are discovered and measured; meanwhile other current champions die.

Wood Densities of Selected Species

Species	Lbs/ft³	Sp. Gr.
Maclura pomifera (Osage-orange)	52.4	0.84
Carya ovata (Shagbark-hickory)	49.9	0.80
Cornus florida (Flowering dogwood)	49.9	0.80
Quercus lyrata (Overcup-oak)	49.9	0.80
Amelanchier arborea (Juneberry)	49.3	0.79
Ostrya virginiana (Hop-hornbeam)	49.3	0.79
Quercus bicolor (Swamp white oak)	49.3	0.79
Carya cordiformis (Bitternut-hickory)	48.7	0.78
Diospyros virginiana (Persimmon)	48.7	0.78
Crataegus punctata (Dotted haw)	47.9	0.77
Carya glabra (Pignut-hickory)	47.4	0.76
Carya tomentosa (Mockernut-hickory)	47.4	0.76
Quercus michauxii (Swamp chestnut-oak)	47.4	0.76
Carya laciniosa (Shellbark-hickory)	46.8	0.75
Oxydendrum arboreum (Sourwood)	46.7	0.75
Ulmus alata (Winged elm)	46.7	0.75
Quercus stellata (Post-oak)	46.2	0.74
Gleditsia aquatica (Water-locust)	45.8	0.73
Morus rubra (Red mulberry)	45.4	0.73
Quercus marilandica (Black-jack oak)	45.4	0.73
Prunus americana (American wild plum)	45.0	0.72
Carpinus caroliniana (Hornbeam)	44.9	0.72
Fraxinus pennsylvanica (Green ash)	44.3	0.71
Quercus alba (White oak)	44.3	0.71
Quercus coccinea (Scarlet oak)	44.3	0.71
Robinia pseudoacacia (Black locust)	44.3	0.71
Malus coronaria (Sweet crab)	43.9	0.70
Gymnocladus dioica (Kentucky coffee-tree)	43.2	0.69
Prunus nigra (Canada-plum)	43.2	0.69
Carya illinoensis (Pecan)	43.1	0.69
Acer saccharum (Sugar-maple)	42.5	0.68
Quercus palustris (Pin-oak)	42.5	0.68
Fagus grandifolia (American beech)	41.8	0.67
Gleditsia triacanthos (Honey-locust)	41.8	0.67
Quercus macrocarpa (Bur-oak)	41.8	0.67
Quercus prinus (Rock chestnut-oak)	41.8	0.67
Quercus velutina (Black oak)	41.8	0.67
Betula lutea (Yellow birch)	41.2	0.66
Quercus rubra (Northern red oak)	41.2	0.66
Ulmus thomasii (Rock-elm)	41.2	0.66
Fraxinus americana (White ash)	40.0	0.64
Cercis canadensis (Redbud)	39.7	0.64
Acer nigrum (Black maple)	38.7	0.62
Quercus falcata (Southern red oak)	38.7	0.62
Betula papyrifera (Paper birch)	37.5	0.60
Fraxinus quadrangulata (Blue ash)	37.5	0.60

Wood Densities of Selected Species *(continued)*

Species	Lbs/ft³	Sp. Gr.
Cladrastis lutea (Yellow-wood)	36.2	0.58
Betula nigra (River birch)	35.6	0.57
Ulmus rubra (Slippery elm)	35.6	0.57
Celtis occidentalis (Northern hackberry)	35.0	0.56
Juglans nigra (Black walnut)	35.0	0.56
Larix laricina (Tamarack)	35.0	0.56
Acer rubrum (Red maple)	34.3	0.55
Betula populifolia (Gray birch)	34.3	0.55
Nyssa sylvatica (Black gum)	34.3	0.55
Ulmus americana (American elm)	34.3	0.55
Platanus occidentalis (American sycamore)	33.7	0.54
Fraxinus nigra (Black ash)	33.1	0.53
Liquidambar styraciflua (Sweet gum)	33.1	0.53
Pinus virginiana (Virginia-pine)	33.1	0.53
Prunus serotina (Black cherry)	33.1	0.53
Magnolia acuminata (Cucumber-tree)	32.5	0.52
Acer saccharinum (Silver-maple)	31.8	0.51
Juniperus virginiana (Eastern red cedar)	30.6	0.49
Acer negundo (Boxelder)	30.0	0.48
Taxodium distichum (Bald cypress)	30.0	0.48
Sassafras albidum (Sassafras)	29.3	0.47
Pinus banksiana (Jack-pine)	28.7	0.46
Castanea dentata (American chestnut)	28.1	0.45
Asimina triloba (Pawpaw)	27.5	0.44
Liriodendron tulipifera (Tulip-tree)	26.8	0.43
Magnolia tripetala (Umbrella-tree)	26.8	0.43
Populus deltoides (Cottonwood)	26.8	0.43
Tsuga canadensis (Eastern hemlock)	26.8	0.43
Catalpa speciosa (Northern catalpa)	26.2	0.42
Prunus pensylvanica (Pin-cherry)	26.2	0.42
Aesculus glabra (Ohio-buckeye)	25.6	0.41
Populus grandidentata (Big-toothed aspen)	25.6	0.41
Salix nigra (Black willow)	25.6	0.41
Populus heterophylla (Swamp-cottonwood)	25.5	0.40
Juglans cinerea (Butternut)	25.0	0.40
Populus tremuloides (Quaking aspen)	25.0	0.40
Tilia americana (Basswood)	25.0	0.40
Aesculus flava (Sweet buckeye)	23.7	0.38
Pinus strobus (Northern white pine)	23.1	0.37
Thuja occidentalis (Eastern arborvitae)	20.0	0.32

Sources: Data for many wood weights are from Romeyn B. Hough, *Handbook of the Trees of the Northern States and Canada* (New York: Macmillan, 1960), 470 p. Others were calculated from specific gravities given by Charles C. Deam, *Trees of Indiana* (Indianapolis: Indiana Department of Conservation, 1953), 330 p.

Note: Wood densities are based on oven-dry weights.

Number of Seeds per Pound and Dispersal Agents for Selected Indiana Tree Species

Species	Average Number of Seeds/Lb[2]	Primary Dispersing Agents
Carya laciniosa (Shellbark hickory)	30	Tree Squirrels
Juglans cinerea (Butternut)	30	Tree Squirrels, Chipmunks
Juglans nigra (Black walnut)	40	Tree Squirrels, Chipmunks
Quercus macrocarpa (Bur oak)	75	Tree and Flying Squirrels
Quercus prinus (Chestnut oak)	75	Tree and Flying Squirrels
Carya tomentosa (Mockernut hickory)	90	Tree Squirrels
Carya ovata (Shagbark hickory)	100	Tree and Flying Squirrels, Chipmunks
Carya illinoensis (Pecan)	100	Tree Squirrels, Chipmunks, Crows
Quercus michauxii (Swamp chestnut oak)	100	Tree and Flying Squirrels, Water
Quercus rubra (Northern Red oak)	105	Tree and Flying Squirrels, Chipmunks
Quercus bicolor (Swamp white oak)	125	Tree Squirrels, Water
Castanea dentata (American chestnut)	130	Tree Squirrels, Birds
Quercus lyrata (Overcup oak)	130	Tree Squirrels, Water
Quercus alba (White oak)	150	Tree and Flying Squirrels, Birds, Chipmunks
Carya cordiformis (Bitternut hickory)	155	Tree Squirrels
Quercus velutina (Black oak)	250	Mammals, Birds
Quercus coccinea (Scarlet oak)	280	Mammals, Birds
Quercus stellata (Post oak)	400	Mammals, Birds
Quercus palustris (Pin oak)	410	Mammals, Birds, Water
Quercus imbricaria (Shingle oak)	415	Mammals, Birds
Acer saccharinum (Silver maple)	1,400	Wind, Water, Mammals
Fagus grandifolia (American beech)	1,600	Mammals, Birds
Nyssa sylvatica (Black gum)	2,000	Birds, Mammals
Gleditsia triacanthos (Honey locust)	2,800	Mammals, Birds
Celtis occidentalis (Hackberry)	4,300	Birds, Mammals
Magnolia acuminata (Cucumber tree)	4,600	Birds, Mammals
Prunus serotina (Wild Black Cherry)	4,800	Birds, Many Mammal Species
Taxodium distichum (Bald cypress)	4,800	Water, Wind
Tilia americana (American basswood)	5,000	Wind, Small Mammals
Acer saccharum (Sugar maple)	6,100	Wind, Mammals
Fraxinus quadrangulata (Blue ash)	6,500	Wind, Mammals
Ulmus thomasii (Rock elm)	7,000	Wind, Small Mammals
Fraxinus nigra (Black ash)	8,100	Wind, Tree Squirrels
Fraxinus americana (White ash)	10,000	Wind, Tree Squirrels
Acer negundo (Boxelder)	11,800	Wind, Small Mammals

Number of Seeds per Pound and Dispersal Agents for Selected Indiana Tree Species *(continued)*

Species	Average Number of Seeds/Lb[2]	Primary Dispersing Agents
Liriodendron tulipifera (Tulip tree)	14,000	Wind, Mammals, Birds
Fraxinus pennsylvanica (Green ash)	17,300	Wind, Tree Squirrels
Acer rubrum (Red maple)	22,500	Wind, Mammals
Robinia pseudoacacia (Black locust)	24,000	Wind, Birds, Mammals
Pinus strobus (Eastern white pine)	27,000	Wind, Small Mammals
Ulmus rubra (Slippery elm)	41,000	Wind, Small Mammals
Juniperus virginiana (Eastern red cedar)	43,200	Birds, Mammals
Ulmus americana (American elm)	68,000	Wind, Small Mammals
Liquidambar styraciflua (Sweetgum)	82,000	Wind, Water
Pinus banksiana (Jack pine)	131,000	Wind, Birds
Tsuga canadensis (Eastern hemlock)	187,000	Wind, Birds
Platanus occidentalis (Sycamore)	204,000	Wind, Water
Larix laricina (Tamarack)	318,000	Wind, Water
Thuja occidentalis (Northern white cedar)	346,000	Wind, Water
Populus deltoides (Eastern cottonwood)	350,000	Wind, Water, Small Mammals
Betula lutea (Yellow birch)	447,000	Wind, Water
Betula papyrifera (Paper birch)	1,380,000	Wind, Water
Salix nigra (Black willow)	2–3 million	Wind, Water, Voles
Populus tremuloides (Quaking aspen)	*ca.* 3,600,000	Wind, Water

[1]Data from most species are from William M. Harlow and Ellwood S. Harrar, *Textbook of Dendrology* (New York: McGraw Hill, 1958), 561 p.

[2]Based on dry weights of seeds only, i.e., without husks, acorn cups, spiny burs, fleshy fruit, wings, or other attachments for fruit/seed dispersal.

Works Cited

Baker, Frederick S. 1950. *Principles of Silviculture.* New York: McGraw-Hill. 414 p.

Bramble, W. C., and F. T. Miller. 1966. "Forestry." Pp. 547–560 in A. A. Lindsey, ed., *Natural Features of Indiana.* Indianapolis: Indiana Academy of Science. 600 p.

Braun, E. L. 1950. *Deciduous Forests of Eastern North America.* New York: Hafner. 596 p.

Cope, Edward A. 2001. *Muenscher's Keys to Woody Plants.* Ithaca, N.Y.: Comstock Publishing Associates. 337 p.

Deam, Charles C. 1931. *Trees of Indiana.* 2nd ed. Indianapolis: Indiana Department of Conservation. 326 p.

———. 1932. *Shrubs of Indiana.* Indianapolis: Indiana Department of Conservation. 380 p.

———. 1940. *Flora of Indiana.* Indianapolis: Indiana Department of Conservation. 1236 p.

Deam, Charles C., and Thomas E. Shaw. 1953. *Trees of Indiana.* 3rd ed. Indianapolis: Indiana Department of Conservation. 330 p.

Elias, Thomas S. 1987. *The Complete Trees of North America: Field Guide and Natural History.* New York, N.Y.: Gramercy. 948 p.

Fowells, H. A. 1965. *Silvics of Forest Trees of the United States.* Agriculture Handbook no. 271, U.S. Forest Service, U.S. Department of Agriculture, Washington, D.C. 20250. 762 p.

Gleason, Henry A., and Arthur Cronquist. 1991. *Manual of Vascular Plants of Northeastern United States and Adjacent Canada.* 2nd ed. Bronx, New York: The New York Botanical Garden. 910 p.

Harlow, William M. 1959. *Fruit Key and Twig Key to Trees and Shrubs.* New York, N.Y.: Dover. 50 + 56 p.

Homoya, M. A., D. B. Abrell, J. R. Aldrich, and T. W. Post. 1985. "The Natural Regions of Indiana." *Proceedings of the Indiana Academy of Science* 94: 235–268.

Hough, Romeyn B. 1960. *Handbook of the Trees of the Northern States and Canada.* New York, N.Y.: Macmillan. 470 p.

Indiana Department of Natural Resources. 2000. *Indiana Big Tree Register 2000.* Indianapolis, Ind.: Division of Forestry, IDNR. 17 p.

Jackson, Marion T. 1969. "Hemmer Woods." Pp. 98–101 in A. A. Lindsey, D. V. Schmelz, and S. A. Nichols, *Natural Areas in Indiana and Their Preservation.* Lafayette: Indiana Natural Areas Survey, Purdue University. 594 p.

―――, ed. 1997. *The Natural Heritage of Indiana.* Bloomington: Indiana University Press. 482 p.

Jensen, R. J., R. De Piero, and B. K. Smith. 1984. "Vegetative characters, population variation and the hybrid origin of *Quercus ellipsoidalis." American Midland Naturalist* 111: 364–370.

Leopold, Donald J. 1998. *Trees of the Central Hardwood Forests of North America: An Identification and Cultivation Guide.* Portland, Ore.: Timber Press. 469 p.

Lindsey, A. A., W. B. Crankshaw, and S. A. Qadir. 1965. "Soil Relations and Distribution Map of the Vegetation of Presettlement Indiana." *Botanical Gazette* 126: 155–163.

Lindsey, Alton A., Damian V. Schmelz, and Stanley A. Nichols. 1969. *Natural Areas in Indiana and Their Preservation.* Lafayette: Indiana Natural Areas Survey, Purdue University. 594 p.

Little, Elbert L. 1953. *Checklist of Native and Naturalized Trees of the United States (including Alaska).* Agriculture Handbook no. 41. Washington, D.C.: U.S. Forest Service. 472 p.

―――. 1980. *The Audubon Society Field Guide to Eastern North American Trees.* New York, N.Y.: Knopf. 714 p.

Melhorn, Wilton N. 1997. "Indiana on Ice: The Late Tertiary and Ice Age History of Indiana Landscapes." Pp. 14–27 in Marion T. Jackson, ed., *The Natural Heritage of Indiana.* Bloomington: Indiana University Press. 482 p.

Mohlenbrock, Robert H. 1973. *Forest Trees of Illinois.* Springfield: Division of Forestry, Illinois Department of Conservation. 178 p.

Newman, James E. 1997. "Our Changing Climate." Pp. 84–99 in Marion T. Jackson, ed., *The Natural Heritage of Indiana.* Bloomington: Indiana University Press. 482 p.

Otis, Charles H. 1931. *Michigan Trees.* Ann Arbor: The University of Michigan Press. 362 p.

Overlease, W. R. 1977. "A Study of the Relationship between Scarlet Oak (*Quercus coccinea* Muench.) and Hill Oak (*Quercus ellipsoidalis* E. J. Hill) in Michigan and Nearby States." *Proceedings of the Pennsylvania Academy* 51: 47–50.

―――. 2001. "Maps of the distribution of certain introduced tree species naturalizing into Indiana." N.p.

Peattie, Donald Culross. 1966. *A Natural History of Trees of Eastern and Central North America.* Boston, Mass.: Houghton Mifflin. 606 p.

Petty, R. O., and M. T. Jackson. 1966. "Plant Communities." Pp. 264–296 in A. A. Lindsey, ed., *Natural Features of Indiana.* Indianapolis: Indiana Academy of Science. 600 p.

Rogers, Julia Ellen. 1905. *The Tree Book: A Popular Guide to a Knowledge of the Trees of North America and to Their Uses and Cultivation.* New York, N.Y.: Doubleday, Page and Co. 589 p.

Rothrock, Paul E. 1997. "The Life and Times of the Tuliptree."
Pp. 257–263 in Marion T. Jackson, ed., *The Natural
Heritage of Indiana.* Bloomington: Indiana University Press.
482 p.

Sargent, Charles S. 1965 [1949]. *Manual of the Trees of North
America.* Vols. 1 and 2. New York: Dover. 934 p.

Shaw, T. E. 1981. *Fifty Trees of Indiana.* West Lafayette: Division
of Forestry, Indiana DNR, and Department of Forestry and
Conservation, Purdue University. 63 p.

Swink, Floyd, and Gerould Wilhelm. 1994. *Plants of the Chicago
Region.* Indianapolis: Indiana Academy of Science. 921 p.

Voss, E. G. 1972, 1985, 1996. Michigan Flora. *A Guide to the
Identification and Occurrence of the Native and Naturalized
Seed-Plants of the State.* 3 parts. Ann Arbor: Cranbrook
Institute of Science Bulletin 59, and University of Michi-
gan Herbarium. Part I, Bulletin 55, 488 p.; Part II, Bulletin
59, 727 p.; Part III, Bulletin 61, 622 p.

Glossary

Acorn. The "nut-like" fruit of an oak tree, not including the cup or stem.

Alternate. Not located opposite to each other on the stem, but each one located at regular intervals, usually on different sides in sequence.

Arcuate. Curved as an arc, usually upward.

Aromatic. Having an odor or aroma, usually pleasant-smelling.

Axis. The central part of a long supporting part of a plant on which plant structures are arranged.

Base. The lower part of a plant structure; with leaves, the portion nearest the leafstalk, or point of attachment. Also adj. basal.

Berry. A fleshy or pulpy fruit with many seeds; it does not open at maturity.

Bisexual. Having both pistils (female parts) and stamens (male parts) present in a given flower.

Blade. The expanded portion of a leaf.

Bloom (noun). A whitish, powdery, or waxy covering of a plant surface such as on a twig or fruit.

Bloom (verb). The process of a flower opening.

Bog. An acidic glacial lake–filled depression containing sphagnum mosses and characteristic plant species growing in the wet spongy conditions.

Bract. Usually a small leaf-like structure attached just below a flower. In flowering dogwood, the bracts are very large, white, showy, and resemble petals.

Branch. In woody plants, any division or subdivision of a stem, except the growth of the season.

Bud. A twig structure that contains the unexpanded leaf, flower, or both. Usually covered with one or more scales.

Canopy. The collective tree crowns of a forest. Canopies may be continuous or discontinuous, closed or with openings.

Capsule. A dry fruit of more than one section that usually opens at maturity.

Catkin. A dry, scaly flower spike, usually unisexual. Typical of willows, birches, walnuts, hickories, oaks, poplars, and some other tree groups.

Chambered. Said of pith which has hollow spaces interrupted by solid partitions.

Compound leaf. A leaf that is further divided into separate, similar leaflets.

Cone. Fruit of the pine and certain other families of coniferous ("evergreen") trees. Usually scaly, often woody, and produces pollen or seeds. Typical of the gymnosperm trees, and certain other families.

Continuum. A continuous series of time, events, conditions, or changes.

Converging. Tending toward one point; if the lines were continued, they would cross, e.g., the lines of the sides of leaf lobes.

Cordate. Heart-shaped; notched at the base and pointed at the tip.

Cork(y). Woody plant material that is lightweight and spongy; made of or like cork.

Crescent. Curved; shaped like a partial moon.

Crown. The top or head of the branches or foliage of a tree.

Deciduous. Dropping or falling away at the close of the growing season. Most trees of Indiana have deciduous leaves.

Diaphragm. A more dense partition within the pith of twigs of certain trees.

Dissected. Cut or divided into several segments.

Divergent. Tending away from one point; if the lines were continued, they would not cross, e.g., the lines of the sides of a leaf lobe.

Divided. Lobed or separated to the base, e.g., leaf lobes separated to the mid-vein of a leaf.

Drupe. A simple one-seeded fleshy or pulpy fruit with a hard and stony seed portion.

Ecological. Concerning the relationship of trees to their environment.

Elliptical (ellipse). Somewhat oval in outline; rounded nearly equally at both ends.

Entire. Having an even edge or margin; not toothed, notched, or divided.

Evergreen. Trees with leaves (either blade-like or needle-like) that persist throughout the year, giving the trees a green appearance throughout their lifespan.

Exotic. An organism or species that is native to or originating from a different country or area from where it currently lives.

Extirpated. Eliminated entirely from a given area, as extirpated from Indiana.

Fen. A low marshy habitat characterized by neutral to basic conditions and supporting wet prairie and other herbaceous plant species.

Fissured. Having openings, channels, or furrows made by the splitting of the bark, especially on very old individuals.

Flatwoods. Forested areas on nearly level land that typically have slow external and/or internal drainage, resulting in seasonal ponding.

Flora. A taxonomic listing of the plant species that occur in a given area.

Floret. The individual flower of a flower cluster or inflorescence.

Flower. A plant reproductive structure bearing stamens or pistils or both.

Fruit. The seed-bearing structure of a plant. It results from the maturing of ovaries after fertilization of the flower.

Furrowed. Alternating ridges and fissures (which see) in the pattern or structure of tree bark. More common on older trees.

Gall. An outgrowth on a plant organ (e.g., a leaf); usually caused by insects, bacteria, or a parasitic fungus.

Genus (pl. Genera). A classification unit next higher than a species, which includes one or more individual species.

Glabrous. Surfaces which are smooth to the touch. Not hairy or roughened.

Glacial relict. A tree species or ecological community that is characteristic of conditions existing during the retreat of the glaciers.

Glade. An open area in a forest, vegetated primarily by herbaceous plants, usually having thin soils over rocky substrates.

Gland. A plant surface or structure that secretes substances; often expanded or extending from the surface of a leaf, stem, or root.

Habitat. The kind of locality or environmental situation in which a tree or plant community grows.

Hairy. A sparse or more dense cover with hair-like structures on plant surfaces such as leaves, buds, or twigs.

Hardwoods. Usually refers to deciduous trees (which see) that have wood densities that typically are harder than those of most coniferous trees (which see).

Heartwood. The dead innermost wood of a tree trunk which is usually harder, darker in color, and more resistant to decay than the outer sapwood.

Herbarium. A place where dried (or otherwise preserved) plant specimens are kept for teaching and scientific study. Also the collection itself.

Hybrid. An intermediate form of plants that originates when two closely related species cross, i.e., one species pollinates another.

Inflorescence. A flower cluster of a plant, and especially its arrangement.

Introduced. Non-native to an area. Brought intentionally from another region, usually for cultivation, as an ornamental, or for establishing a tree plantation.

Key. An abbreviated set of contrasting diagnostic characters for identifying trees or other organisms.

Ladder-like. Pinnate, i.e., having leaves divided into leaflets or segments usually arranged oppositely along a common axis or leaf stem.

Lanceolate. Lance-shaped, much longer than wide.

Leader. The terminal or topmost shoot of a tree.

Leaf. The expanded blade-like plant organ that is the site of photosynthesis or food production.

Leaflet. One of the divisions of a compound leaf.

Leafstalk. A leaf petiole, i.e., the support structure that attaches a leaf to a twig. Also, as used here, the central stem of a compound leaf.

Legume. The dry fruit of plants of the Fabaceae (Leguminosae) that opens at maturity along two sides. Loosely, a "pod."

Lenticel. Small lens-shaped corky growths along the twigs of many tree species. Loosely referred to as "pores" herein.

Lobe. Any segment of a plant organ, especially if rounded. Usually referred to are lobes of leaves or flower petals.

Lustrous. Having luster or sheen, shining.

Mid-rib (mid-vein). The central rib or vein of a leaf, petal, or other plant organ.

Native. Occurs naturally in the area where it is found, indigenous; not introduced from elsewhere.

Naturalized. Not originally native to the region where found, but so well established as to have become a part of the flora. Introduced successfully.

Needle. A leaf structure that is long, linear, or very narrow. Often round or half-round in cross-section and evergreen.

Niche. The specific ecological position a plant species occupies in nature; where it lives and what role it performs.

Nut. A hard, often woody, one-seeded fruit that typically does not open at maturity. Does not include the surrounding layers or support structure of the fruit.

Obovate. Inversely ovate (which see).

Opposite. Leaves, branches, or buds which appear directly across from one another on a plant stem.

Oval. Broadly elliptical.

Ovary. The part of the female flower which contains the ovules (which see), and which matures into the fruit.

Ovate. Egg-shaped; having the outline like that of an egg with the broader end basal.

Ovule. The flower structure which after fertilization develops into the seed.

Palmate. Diverging radially like the extended fingers of the human hand; hand-like.

Petal. One of the divisions of the often showy, non-reproductive, leaf-like flower structures. A segment of a "bloom."

Petiole. The leafstalk; the support stem of a leaf.

Pinnate. Having leaves divided into leaflets or segments along a common axis. See ladder-like.

Pistil. The female part(s) of a flower, usually differentiated into ovary, style, and stigma.

Pith. The spongy center tissue of a twig, chiefly consisting of soft cells with unthickened walls.

Pod. A dry, often elongated, fruit that usually opens at maturity.

Pore. An opening into a plant surface structure. As used herein, a lenticel (which see).

Prickle. A sharp outgrowth from the outer covering of a plant.

Putative. Commonly thought or supposed; reputed to be.

Recurved. Curved downward or backward, e.g., recurved teeth on a leaf edge.

Remotely. Barely or only slightly, e.g., remotely toothed margin or edge.

Resinous. Sticky plant sap, especially of coniferous trees, that sometimes oozes from wounds or surfaces.

Root. The underground part of a plant which supplies it with water and nourishment.

Samara. A thin, flat-winged dry fruit with a single seed that typically does not open at maturity, as in the ashes, maples, elms, and other groups.

Sapwood. The younger, often-living, outermost wood of a tree trunk which is typically softer, more moist, lighter in color, and more susceptible to decay than heartwood.

Scales. The outermost, minute leaf-like structures that cover plant buds.

Scaly. A type of bark pattern in which thin layers separate along one edge, yet remain attached at the other.

Seed. The ripened ovule consisting of the embryo, stored food, plus any surrounding essential coverings. In the plural, all such structures of the same species.

Serrate. Having sharp teeth pointing forward, as typically occur on a saw.

Sessile. Without a stalk; said of leaves, leaflets, buds, or flowers.

Shrub. A woody perennial, smaller than a tree, usually less than 10 feet tall and with several stems.

Silky. Covered with closely pressed, soft, and straight hairs.

Sinus. The opening or space between two lobes of a leaf.

Site. The habitat conditions where a tree is growing. It involves topographic, soil, and microclimate conditions, plus the effects of associated plant species.

Slough. A depression that is filled with water for at least part of the growth season.

Softwoods. Usually refers to coniferous tree species collectively. Their woods are typically less dense than deciduous trees, but not universally so.

Species. A group of individual plants that reproduce among themselves in nature, and which are reproductively isolated and distinguishable from other such groups.

Spherical. Globe-shaped; a round plant structure, such as an acorn, that has all radii of approximately an equal length.

Spine. A sharp woody outgrowth that is not a modified branch, i.e., a sharpened leaf, stipule, or epidermal structure.

Stamen. The male or pollen-bearing part(s) of a flower, normally consisting of an anther and a filament.

Stem. The main ascending axis of a plant.

Stipule. A small structure, frequently leaf-like, at the base of a leafstalk.

Stool. A cut or burned-off tree stump at ground level which gives rise to multiple shoots. Also verb, to regrow from such a stump, e.g., to "stool out."

Storax. An aromatic resin or balsam derived from certain trees or shrubs, e.g., Sweet gum.

Succession. The process of rebuilding ecological communities following their complete removal or extensive disturbance.

Swamp. A forested wetland. May be located along a stream course or within an upland depression. May also be vegetated with shrubs, i.e., a shrub swamp.

Taxon (pl. Taxa). A unit of classification, regardless of rank, e.g., a family, a genus, or a species.

Taxonomy. The science of classifying organisms.

Thorn. A sharp woody structure that is a modified branch. Thorns have vascular tissue; spines do not.

Tolerant. Capable of enduring more or less heavy shade.

Toothed. Having regularly spaced, notched indentations along the margin of a leaf or petal.

Topography. The structure of the landscape; it varies from level to steep, and from uniform to strongly dissected.

Tree. Perennial woody plant, usually with a single stem, an evident trunk, and typically greater than 10 feet tall.

Truncated. Ending abruptly as if cut off, e.g., the squared-off tip of a Tulip-tree leaf.

Undergrowth. The plants collectively of the ground layer within a forest. Sometimes the shrub layer vegetation is included as well.

Understory. The subcanopy trees of a forest considered as a unit.

Unisexual. Flower of one sex, either staminate (male) or pistillate (female) only.

Variety. A subdivision of a plant species; closely synonymous with a subspecies.

Vascular tissue. Specialized conducting cells which carry water and solutes upward from the roots and throughout a plant, and carry food materials downward from the leaves and throughout the plant.

Vegetation. The plants collectively that make up the plant community of a region. The plants as a structural unit as opposed to a list of species present (see flora).

Vein. The thread of vascular tissue in a leaf or other plant organ.

Whorl. An arrangement of leaves, etc. in a circle around the stem.

Wing. Any membranous or thin expansion bordering or surrounding a twig, seed, fruit, pollen grain, or other plant organ.

Woolly. Clothed with long, matted, dense hairs.

English–Metric Measurements Conversion

Converting English Units to Metric Units

Measure	English Unit	Metric Unit
Weight	1 pound	0.45 kilogram
	1 ounce	28.35 grams
Length	1 mile	1.61 kilometers
	1 yard	0.91 meter
	1 foot	0.30 meter
	1 inch	2.54 centimeters
Square Measure	1 acre	0.40 hectare
	1 square mile	2.59 square kilometers

Converting Metric Units to English Units

Measure	Metric Unit	English Unit
Weight	1 kilogram	2.20 pounds
	1 gram	0.04 ounce
Length	1 kilometer	0.62 mile
	1 meter	3.28 feet
	1 centimeter	0.39 inch
Square Measure	1 hectare	2.47 acres
	1 square kilometer	0.39 square mile

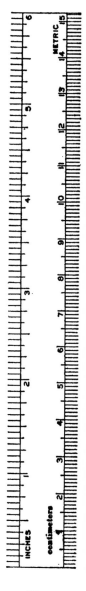

Index

Species that are excluded, extirpated, or subject to taxonomic revision are indicated by an asterisk (*).

KATHERINE HARRINGTON earned her B.F.A. in graphic design at Indiana State University. She is a graphic artist at Hindostone Products in Indianapolis.

MARION T. JACKSON earned a B.S. in Conservation of Natural Resources and a Ph.D. in Plant Ecology from Purdue University. He is Professor Emeritus of Ecology at Indiana State University, former Chairman of the Indiana Chapter, The Nature Conservancy, and Past President of the Indiana Academy of Science. He is editor of *The Natural Heritage of Indiana* (Indiana University Press, 1997).

RON RATHFON is a forester for Purdue University's Department of Forestry and Natural Resources. He is stationed in Dubois County, where he conducts applied forestry research and manages the forest. His photography reflects his lifelong passion for trees.